fP

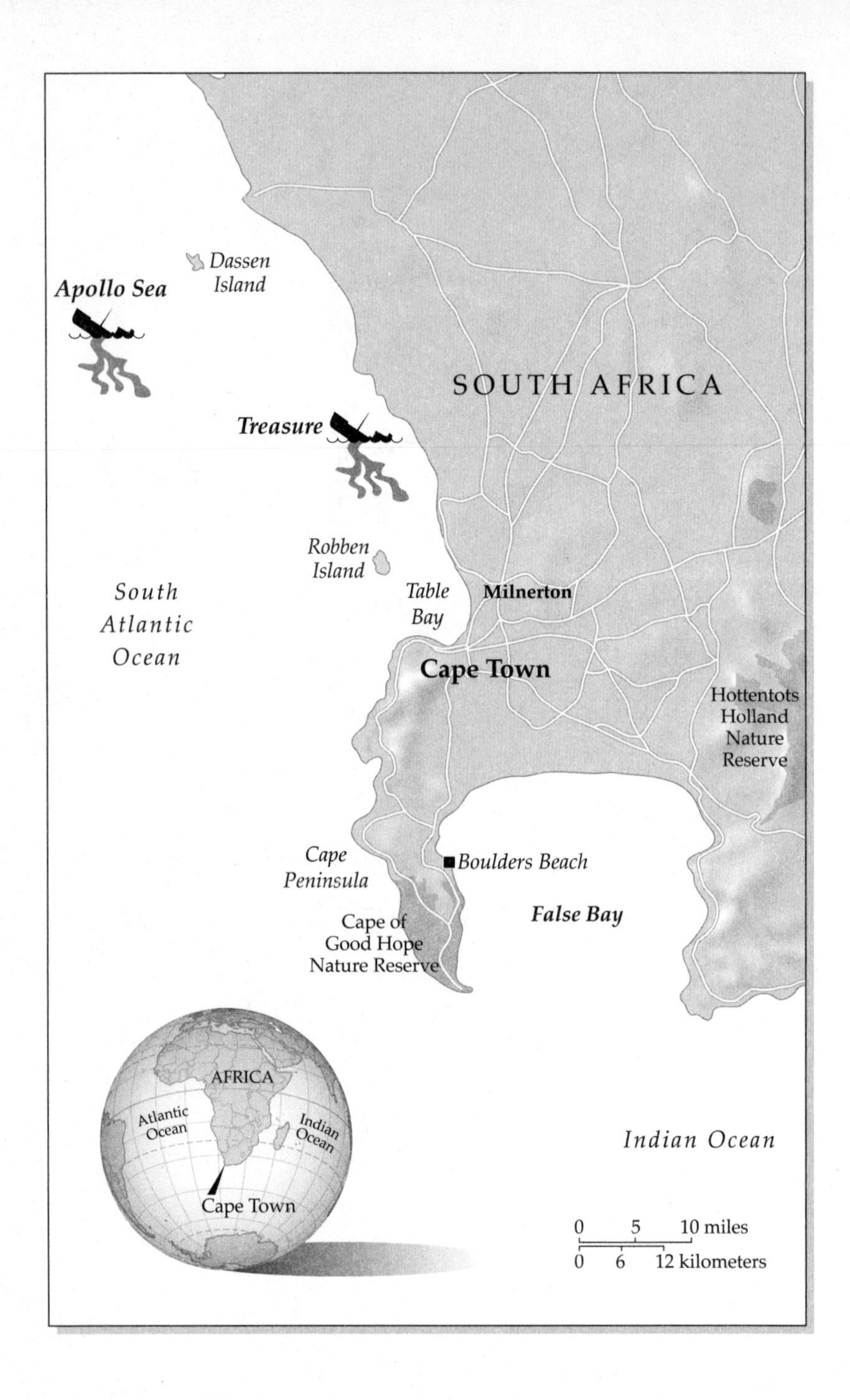

Locations between Dassen and Robben Islands where the iron-ore carriers *Treasure* and *Apollo Sea* sank and spilled their oil. The two ships sank exactly six years and three days apart.

THE GREAT PENGUIN RESCUE

40,000 Penguins, a Devastating Oil Spill, and the Inspiring Story of the World's Largest Animal Rescue

DYAN DeNAPOLI

Free Press
New York London Toronto Sydney

The author is donating a combined total of 20 percent of her advance and any additional royalties from this book (net of agency commissions and taxes) to one or more of the fifteen rescue, research, and conservation organizations listed under one of the Penguin headings in Appendix II, as well as to organizations devoted to the Gulf oil spill rescue efforts.

fP

Free Press
A Division of Simon & Schuster, Inc.
1230 Avenue of the Americas
New York, NY 10020

First Free Press hardcover edition October 2010

FREE PRESS and colophon are trademarks of Simon & Schuster, Inc.

For information about special discounts for bulk purchases, please contact Simon & Schuster Special Sales at 1-866-506-1949 or business@simonandschuster.com.

The Simon & Schuster Speakers Bureau can bring authors to your live event. For more information or to book an event contact the Simon & Schuster Speakers Bureau at 1-866-248-3049 or visit our website at www.simonspeakers.com.

Designed by Jill Putorti

Manufactured in the United States of America

10 9 8 7 6 5 4 3 2 1

Library of Congress Cataloging-in-Publication Data

DeNapoli, Dyan.
The great penguin rescue : 40,000 penguins, a devastating oil spill, and the inspiring story of the world's largest animal rescue / Dyan deNapoli.
p. cm.
1. African penguin–Effect of oil spills on–South Africa–Atlantic Coast. 2. Oil spills and wildlife–South Africa–Atlantic Coast. 3. Wildlife rescue–South Africa–Cape Town Region. 4. Wildlife rehabilitation–South Africa–Cape Town Region. 5. DeNapoli, Dyan. I. Title.
QL696.S473D46 2010
639.9'78470968–dc22 2010017156

ISBN 978-1-4391-4817-4
ISBN 978-1-4391-5486-1 (ebook)

Maps by Paul J. Pugliese.

This book is dedicated to my extraordinary parents, Phyllis Carter deNapoli and Paul Angelo deNapoli, who have always loved, supported, and guided me. They were shining examples of how to treat others and how to live a meaningful life. They believed in me–long before I learned to believe in myself–and I am who I am because of them. I am profoundly grateful to my remarkable mother; although she is now in spirit, I still feel her loving presence every day.

And to every person who worked so tirelessly to save the penguins from the *Treasure* oil spill: thank you, from the bottom of my heart.

Contents

THE GREAT PENGUIN RESCUE

PROLOGUE:

Black Waters—Panic at Sea

The question is not, "Can they reason?" nor, "Can they talk?" but rather, "Can they suffer?"

—JEREMY BENTHAM, EIGHTEENTH-CENTURY PHILOSOPHER

There they were. The scales on the sardines flashed and shimmered as they reflected the sunlight streaming through the water. After feeding their ravenous chicks for two straight days, and having swum several miles to reach the foraging grounds, the penguins were ready to eat. While they usually went out to sea in small groups, once they located a school of fish, every penguin had to isolate and capture their own prey. Each bird was now on its own. One of the penguins took a deep breath and dove beneath the sparkling surface of the ocean, swimming until it was below the schooling fish. The penguin hovered there, its black back blending in with the dark ocean floor, helping to conceal it from the sardines above. Then, in a sudden burst of speed, it shot up through the swirling mass, grasped a silvery fish behind its gills, and, while still underwater, swallowed it headfirst and whole. A swift and agile hunter, the penguin caught and swallowed several more fish before its aching lungs signaled the need to come up for air. After being underwater for several minutes, it surfaced far from where it had originally submerged.

Only now, the penguin found itself in the midst of a thick and noxious substance that clung to its feathers and slowed it down as it swam. The caustic oil got into the bird's eyes, burning them and making it hard to see. Confused and anxious, the penguin struggled to make its way through the viscous black stuff floating on the surface of the ocean. The heavy oil coating its body weighed it down, making it hard to keep its head above water. The penguin frantically pumped its wings, but it was becoming increasingly difficult to move. With every breath, it inhaled some water, along with the traces of oil coating its beak. Choking on the toxic mix burning its lungs and throat, the penguin coughed and struggled to breathe.

The sticky oil had caused the penguin's dense, overlapping feathers to clump and separate, and the cold ocean waters now penetrated its feathers like icy fingers. The water eventually reached the penguin's skin; as its body temperature plummeted and hypothermia set in, it became weak and disoriented. The penguin swung its head from side to side, searching for the nearest landmass. If it could make it to shore, it might get some relief from the cold and the fumes. There was an island several miles off in the distance, but did the penguin have the strength to swim that far? Instinct drove it to head in that direction. But, in its weakened state, it was several strenuous hours before the island was within reach. As the penguin made its final approach, the breaking waves tossed it violently against the rocks, which were now slick with oil, causing it to slip and struggle to get its footing. Exhausted, the penguin finally heaved itself onto the rocky beach, where hundreds of other penguins stood huddled together, the heavy black oil that slowly dripped from their bodies forming expanding black puddles around their feet.

Some of the penguins stood statue-still. Hunched over, their wings hanging limply by their sides, they were in a state of shock. Others were compulsively preening themselves, trying to remove the thick substance from their bodies; but it was an impossible task. The oil clung to every last feather and, while using their beaks in their futile attempts to clean and straighten them out, the birds were inadvertently swallowing large amounts of the toxic mess. If the oil remained in their intestinal tracts, bleeding ulcers would form, causing their normally green and white feces to turn dark brown from digested blood and swallowed oil. Over time, the toxins from the oil would get into their bloodstreams, where they would break down the red blood cells, leading to anemia. Eventually, the ingested oil could kill them.

The penguins were now landlocked. They could not return to sea to hunt for food, because their soiled feathers no longer provided protection from the icy waters. Any oil-coated penguins that were eventually driven by hunger to brave the waters to feed were quickly forced back to shore by the penetrating cold. Even though schools of fish were just yards away in the ocean, the penguins were compelled to stay on dry land; but standing there, they would soon starve to death. Their hungry chicks would starve as well. It was an impossible situation. There were no good options for the penguins–or their chicks–and there seemed to be no way out of their deadly predicament.

At first, a few hundred penguins were standing on the beaches, then a few thousand, and later, more than 10,000. And still they kept arriving until, in the end, nearly 20,000 penguins covered with oil lined the coasts of South Africa's Robben and Dassen Islands. Those penguins that couldn't make it back to land swiftly enough after swimming through the oil succumbed to hypothermia or drowned. The penguin that had struggled to get there from the feeding grounds that day had been fortunate enough to make it back to shore before meeting either of those fates. Eventually, though, one out of every ten oiled penguins standing there would die.

Would this penguin be one of ten as well? Would anyone come to its rescue?

1

Wing to Wing—20,000 Oiled Penguins

Not to hurt our humble brethren is our first duty to them, but to stop there is not enough. We have a higher mission—to be of service to them wherever they require it.

—ST. FRANCIS OF ASSISI, PATRON SAINT OF ANIMALS

It was early—too early. I was sitting in the back of a bouncing white minibus, trying to get my bearings through a fog of jet lag and exhaustion. At six thirty on a cool winter morning I had been in South Africa for only eight hours, most of which had been spent sleeping after the seemingly endless twenty-seven-hour journey from Boston. Just eight days earlier, on June 23, 2000, a Greek-owned iron-ore carrier had foundered off the coast of Cape Town, polluting the habitat of more than 75,000 African penguins with oil as it sank. The offending ship—ironically, named the *Treasure*—had gone down between two of the penguins' main breeding colonies, on Dassen Island and Robben Island, putting nearly half of the entire world population of this already vulnerable species at risk. Within hours of the spill, thousands of heavily oiled penguins had begun streaming onto their islands in an attempt to escape the icy waters of the southern Atlantic Ocean. Due to the waterproof barrier created by their tightly overlapping feathers, these hardy seabirds can normally survive in the frigid currents that supply them with a bountiful food source for days or even weeks on end. With thousands of small feathers that lie over each other like shingles on a roof, and with microscopic barbules on those feathers that interlock to form an impervious shield, penguins are well equipped for life at sea. But now, the clumping effects of the oil had rendered their feathers useless.

Knowing that an oil-soaked bird has no chance of survival without human intervention, and can last only a matter of days on its own in the wild before it dies, local rescue groups and conservation officials immediately launched a massive effort to capture and rehabilitate the oiled penguins. Seven other penguin professionals and I had just arrived to help with the rescue operation, and we were now on our way to a warehouse two miles from the heart of Cape Town where thousands of oil-covered penguins were being sheltered. More than 16,000 oiled penguins had already been rescued from their breeding islands, and thousands more contaminated birds had yet to be captured. There were so many that they had quickly overrun the local rehabilitation center, and an emergency rescue facility had been constructed to house them. In addition to these birds, another 60,000–that had so far managed to escape being oiled–remained on their islands and would likely come into contact with the drifting oil slick if not removed from their colonies. For the oil-soaked penguins that had been recovered, the rescuers now had to begin the daunting process of washing, feeding, and nursing them all back to health.

Our band of eight, having just convened at JFK Airport in New York the day before, was the first team of zoo and aquarium professionals to arrive from the United States to assist the local rehabilitation center with this enormous effort. We had been summoned to Cape Town because of our hands-on experience with, and extensive firsthand knowledge about, penguins. The members of our group, all hailing from institutions accredited by the Association of Zoos and Aquariums (AZA), had varying amounts of experience, but most had worked closely with penguins for ten to fifteen years. I had been working with penguins at Boston's New England Aquarium for five years; after serving as an intern and volunteer for nearly two years, I was hired as a penguin aquarist in 1997. My co-worker from the aquarium, Heather Urquhart, was also part of the team heading to Cape Town. A senior penguin aquarist at the time of the oil spill, she had been working with penguins for twelve years. Our days were normally spent inside the aquarium's spacious penguin exhibit, caring for a thriving colony of sixty-nine birds. Along with a third staff member (and assistance from several volunteers), we fed the penguins, monitored their health, cleaned their exhibit, raised their chicks, and educated our visitors about them. The other members of our newly formed team all performed similar duties at their institutions. Now, we were about to have our first encounter with the oiled penguins, and we had no idea what to expect or exactly what would be expected of us. We only knew that the size and scope of this rescue operation was unlike anything ever undertaken before. We had been thrust into the midst of this crucial mission with just two days' notice, and with the expectation that we would provide the expertise they urgently needed. Although I tried to remain optimistic, I secretly harbored some doubts about our ability to handle the monstrous challenge that lay ahead.

It was winter in the Southern Hemisphere, and the darkness of the South African morning surrounded us as we made our way to the rescue center. Our group was quiet during the twenty-five-minute ride, each person lost in their own thoughts about the task we were about to face. Although collectively we had more than one hundred years of experience working with penguins, none of us had dealt with a situation of this magnitude. The truth was that no one ever had. In the history of organized wildlife rescue, there had never been this many penguins–or any other kind of animal–oiled and recovered alive at once before. In fact, this penguin rescue would soon prove to be twice as large as any that had been attempted in the past. And it would double again in size before it was over.

Given the astronomical number of animals that had to be rehabilitated, the likelihood of saving most of the penguins seemed doubtful. We were all aware that, in the last large-scale rescue of African penguins six years earlier, following the sinking of the *Apollo Sea* near their largest breeding colony on Dassen Island, more than half of the 10,000 oiled penguins brought in for rehabilitation had died. Apparently, I was not the only one concerned about the odds of a successful outcome for this effort. I later learned that many of my team members had similar doubts; yet none of us voiced our reservations at the time. I'm sure we all wanted to remain strong and positive for each other, for ourselves, and for the penguins. To do so, we had to enter into this endeavor with self-assured, can-do attitudes. If there was ever a moment in each of our lives when we had to pull from a deep, untapped well of strength and determination, this was it. In spite of the staggering circumstances, this was no time to let fear or uncertainty overwhelm us.

As we approached our destination, Cape Town's famous Table Mountain emerged from the darkened skies, dominating our view of the landscape. It was easy to see how this grand sandstone edifice sheltering the city and Table Bay Harbour came by its unusual name–most of its peak appeared to have been sheared clean off with the swipe of an enormous sword blade, leaving in its wake an expansive plateau two miles across. Nestled at the base of this mountain was the temporary rehabilitation center that had been hastily constructed just days before in a railway warehouse. Located in Salt River, an industrial suburb of Cape Town, the makeshift shelter was dubbed the Salt River Penguin Crisis Centre.

An orange glow had just started to spread across the horizon as we passed through the entrance gate and pulled up alongside the building where the penguins were being held. I stepped out of the minibus and started walking toward the enormous steel structure ahead, my pulse quickening as I anticipated what we would find inside. Even at this early hour, with daylight just breaking over Cape Town, the rescue center was already buzzing with activity. Resembling swarming bees tending to a damaged hive, hundreds of frenzied workers dressed in yellow foul weather gear were bustling about in front of the building. The sheer number of volunteers was stunning–and unexpected. There was no discernible organization or process to their chaotic movements, and we watched in awe as they rushed in and out of the enormous warehouse doors. It was impossible to see inside from where we stood, but the dark entryway loomed like a huge yawning mouth waiting to swallow us whole.

I had not yet been lucky enough to visit a penguin colony in the wild, but having read that these sprawling seabird gatherings can be heard (and smelled) long before they can be seen, I fully expected to be greeted by a cacophony of braying and honking upon entering the rescue center. After working with African penguins for several years at the New England Aquarium, I knew firsthand that they were indeed very vocal birds, prone to extended fits of raucous, competitive braying during territorial displays and pair-bonding rituals. Their loud *hee-haw* calls echoed off the cement walls of the aquarium, filling the building with barnyard sounds, much to the surprise of our visitors. As their calls are remarkably similar to the braying sounds made by donkeys, African penguins are sometimes called Jackass penguins or–in South Africa–Beach Donkeys.

But instead of hearing their harsh brays as we stepped through the cavernous doors of the warehouse and into the shadowy interior, we were met with an eerie silence, immediately signaling to us in undeniable terms the stressed mental and physical state of the penguins inside. In a space that should have been reverberating with the boisterous calls of thousands of penguins, the air was heavy and still. The silence itself was like an unearthly presence filling the building. I stood rooted in place, trying to detect any of the usual sounds that should have been flooding the space: the clamor of penguins honking and braying, fighting over territory, and displaying for or calling to a mate. It was the middle of their breeding season, but in the chaos of removing the penguins from the islands, thousands of mated pairs had been abruptly separated from each other. Every one of these penguins had been rudely ripped from their nests, their mates, and their chicks, then tossed haphazardly into random holding pens in the vast warehouse. Standing there, just inside the entrance, I kept waiting to hear the plaintive voices of displaced and lonesome penguins calling out, trying to locate their mates. But the traumatized birds remained mute.

What stunned me even more than the unexpected silence was the nearly physical wall of odor that assaulted us as we crossed the threshold. The stench was horrific. I came to learn that the nauseating odor permeating the air was a combination of acidic guano (penguin excrement), the oil covering the penguins, the sardines being fed to them, coal dust, pungent human sweat, and the food being prepared for the volunteer workforce. The stench was so overwhelming that I was forced to breathe through my mouth to keep from gagging. Every so often, I tried to breathe normally, but as soon as the putrid smell hit my nostrils, I started choking and gagging again. It was a relief when, after an hour or so, my senses adjusted and I was finally able to breathe through my nose without having the constant urge to retch.

Shortly after entering the building we were greeted by Jay Holcomb and Linda Elliott, both from the California-based International Bird Rescue Research Center (IBRRC), and Sarah Scarth, South African director of the International Fund for Animal Welfare (IFAW)–who introduced themselves as our team leaders and the directors overseeing the rescue operation. Internationally recognized as the leading authority in the field of oiled wildlife rehabilitation, the IBRRC has made it its mission to rescue animals in distress for nearly forty years. Since 1971, its workers have responded to more than one hundred fifty oil spills, saving tens of thousands of oiled birds, as well as many mammals and reptiles, throughout the world.

As executive director of the IBRRC and a thirty-five-year veteran of wildlife rescue, Jay Holcomb had been on virtually every rescue mission since joining the organization in 1986. Barrel-chested and short in stature, Jay welcomed us with a warm smile. While he had an easygoing manner, it was immediately clear that he was a take-charge kind of guy. Linda Elliott had been with the IBRRC for five years and was their Hawaii and Pacific Islands field representative. Tanned, with long blond-streaked hair that flowed past her shoulders, she looked as if she had just stepped off a surfboard and into this nightmare. This was her twelfth oil spill response in five years. Sarah Scarth had overseen ten rescue missions during the six years she had been with IFAW. A tall brunette with closely cropped hair and a beautiful smile, she graciously welcomed us to Cape Town, and thanked us for coming to help them. The International Fund for Animal Welfare has been saving wild and domestic animals from natural and human-created disasters worldwide since 1969, and was the umbrella group under which all of the teams worked at the Salt River station. While IFAW managed most of the complex logistics, the IBRRC oversaw the washing and rehabilitation of the penguins.

The *Treasure* was the second large-scale oil spill response for Sarah, Jay, and Linda this year. Just five months earlier, all three had been involved in the international effort to save more than 15,000 oiled seabirds that were rescued after an Italian vessel, the *Erika,* sank off the coast of France. Tragically, almost all of those birds, plus an estimated 150,000–300,000 others that were not recovered, perished in that spill. Most of the oiled birds suffered horrible deaths at sea or on shore before workers could even get to them. Of the 15,000 birds that were brought in alive to the rescue centers, only 1,900–a mere 12.6 percent–survived. The astronomical mortality rate was not for lack of experience or effort on the part of the rescuers–the extent of the spill was just too vast, the number of affected animals too overwhelming, and the degree of their oiling too great. With the experience of that devastating setback and massive loss of life so recently behind them, I could only imagine the intense pressure they must have felt to prevent a repeat of that demoralizing event from occurring here; but their easy smiles and cheerful demeanors belied any stress they might have been feeling. With so many disaster responses under their belts, they kept a laser-sharp focus on what had to be done, and understood the utter importance of keeping the morale and enthusiasm of their staff and volunteers up.

I was immediately in awe of these three and their teams, all of whom live in a constant state of disaster preparedness. Their suitcases are always packed and ready, so that the moment they receive a call about animals in distress, they can head directly to the airport and be on the next available flight. That was exactly what they had all done upon getting the urgent call from Estelle van der Merwe, centre manager of the Southern African Foundation for the Conservation of Coastal Birds (better known as SANCCOB), the local rehabilitation center in Table View, a suburb of Cape Town.

After we had been introduced to the other members of the crisis management team (which included IFAW staff, IBRRC staff, and a few other bird experts) and given a brief tour of the facility, Jay and Linda escorted us to an enormous room that held rows upon rows of round, blue pools, each three feet high and ten feet in diameter. Each pool consisted of a long strip of blue vinyl that was wrapped around a circular chicken-wire frame. This structure was placed on top of a round piece of vinyl on the floor; perforated rubber matting (called Dri-Dek) had been laid down inside the pools to give the penguins a cushioned surface to stand on. The Dri-Dek also allowed some of their guano to drain through, so the birds weren't standing and lying in thick puddles of their own excrement. These looked like the type of pools that would normally be in someone's backyard, filled with water for children to splash around in on a hot summer day. Here, however, they served as dry holding pens for the penguins. When I peered over the side of the pool closest to me, I was astonished to see approximately a hundred penguins huddled together inside.

While penguins are colonial birds, preferring to gather in large groups, they are at the same time extremely territorial and will aggressively defend their personal space with vicious bites and wing slaps. These penguins, though, stood shoulder-to-shoulder and statue-still in the pools, apparently in a state of shock. Their bodies were covered with varying amounts of thick black oil, and now and then, a penguin would begin preening in a vain attempt to clean the viscous substance from its feathers. It was impossible for the birds to remove the oil, but the natural instinct to clean and straighten their feathers occasionally prevailed and they would make another halfhearted attempt before giving up again.

The spectacle of thousands upon thousands of penguins in such a pathetic physical and mental state was heartbreaking. It was hard to keep from breaking down at the sight of so many dazed and suffering birds, but we did not have the luxury of giving in to painful emotions. Like medics or soldiers, we knew that we could not allow our feelings to overwhelm us or get in the way of the work we had come there to do. Surveying my surroundings, I realized there were more than fifty pools filled with oiled penguins in this room alone, and there were three other similar holding rooms in the massive building. I was beginning to grasp the scope of this endeavor, and what 16,000 penguins under one roof actually looked like. I silently wondered just how long it would take to clean them all, nurse them back to health, and get them back out into the wild. Would such a monumental task even be humanly possible?

My inner musings were interrupted by Jay, who directed our attention to one of the holding pens. As our team of experts huddled around one pool brimming with penguins, another team inside the enclosure demonstrated the technique they were using to feed the birds. Because these wild penguins typically caught and ate live, wriggling fish underwater and had no concept of how to take a dead, stiff fish from a person's hands and swallow it, each bird had to be individually force-fed. Jay told us that it usually took two to three hours for a group of three or four people to feed all of the penguins in one pool. After watching the routine for a short time, he instructed the members of our team to enter two of the crowded pens and begin feeding the birds. Even though we all had previous experience force-feeding the occasional sick penguin at our respective institutions, we quickly learned that force-feeding a wild penguin was another matter altogether.

First, you had to catch and restrain one of the penguins. This was done by grabbing a bird by the back of the head, holding it firmly with three fingers on top of the skull—the rear portion of the lower jaw grasped tightly between pinky and thumb—then slipping the other arm under the belly to lift and support the bird. To those who have never had to catch a penguin, this may sound fairly straightforward. In reality, trying to capture and hang on to a frightened wild penguin whose body is slippery with oil, and who is lashing out at your face and body with a hooked beak, is extremely challenging.

Penguins are surprisingly quick and agile and, despite their small size, far stronger than they appear.

And while they may look like neckless wonders–their heads appearing to be attached directly to their shoulders–in truth, they have deceptively long necks, which they can twist and turn more than 180 degrees to attack with the speed of a striking snake. When penguins attack, their S-shaped necks act like coiled springs which they release explosively, darting their head forward in a flash to extend their reach. These long, tightly curved necks are an effective adaptation for snatching darting fish underwater and a fine defense against predators; however, it makes trying to capture them an intimidating and hazardous task. This is why it's advisable to catch a penguin by the back of its head, if at all possible, because the only way to safely control the bird is to have control of its head. If you let go of the wildly swinging head you can be absolutely certain that you–or someone standing close to you–will be badly bitten.

You must not only be very swift in your attempt, you must also be very sneaky about it, because if you make the mistake of looking directly at the penguin you are trying to catch, it will instantly realize your intentions and do everything in its power to evade capture. If you finally do manage to corner a bird, it will go into attack mode and you will surely be on the losing end of that battle. Not only are African penguins' beaks razor-sharp along the edges; they have a pointed hook at the end of the top beak which fits into a slotted groove at the end of the bottom beak. So, when they bite, they slice into your flesh with the sharp edges of their beak while pinching your skin tightly between the hooked ends. Once they have a firm grip on your hand, your thigh or–heaven forbid–your face, they twist their head sharply, leaving not only a bloody cut in the shape of a long V but also a painful welt and bruise in their wake. Even through jeans and thick clothing, their powerful bite can tear skin and draw blood. So, the best approach is to sneak up on an unsuspecting penguin from behind and swoop down to grab the back of its head in one quick stroke.

Unfortunately, a sneak attack with the penguins at Salt River was out of the question because the entire group would huddle tightly together—seemingly trying to disappear through the wall of the pool farthest from where we stood—keeping their sharp beaks pointed directly at us at all times. And since the pools were round, it was impossible to corner them. So humans and penguins would chase each other around the pool in a ridiculous circular dance until one of the humans, resigned to sacrificing arms and hands to a penguin's crushing bite, would make a courageous lunge and grab a bird face-on. I soon noticed that many of the local volunteers were employing an unusual catching method I hadn't seen before. I couldn't decide if it was brave or foolish—or perhaps a bit of both. But they first grasped a penguin by the tip of one wing and gently pulled the bird toward them, and then—using their other hand—grabbed the bird by the back of its head. Knowing the full range of a penguin's freakishly long neck, and recognizing that my hand at the end of their wing would be well within their bite zone, I did not attempt this particular method of capturing them. Instead, I relied on my years of penguin catching at the New England Aquarium and stuck with the technique I had plenty of experience with.

Added to the challenge of catching the penguins was the fact that the oil covering their bodies made it very difficult to hold them firmly against the slick foul weather gear we wore (called oilskins in South Africa). As a result, the birds sometimes slipped out of our grasp the way a bar of soap might fly out of one's hands in the shower. The mayhem of simply trying to catch a penguin often led us to feel as though we were in a Three Stooges or Keystone Kops slapstick routine.

Once we finally managed to catch one of the penguins, we sat on a low stool inside the pool and held the bird in place between our knees by squeezing it tightly beneath its wings—with its back facing our stomach and the weaponlike beak pointed away from us. Only after we had gained control of the bird could we begin to force-feed it. This was done by prying the beak open with both hands (which were thickly gloved for protection), holding the beak open with one hand while reaching down to grab a sardine out of a bucket with the other, then placing the slippery fish headfirst into the penguin's mouth. Finally, we carefully pushed the sardine down the penguin's throat, being sure to direct the fish into the esophagus and not the trachea, where it would block the passage of air and suffocate the bird. If the penguin cooperated, it eventually started to swallow the fish, slowly taking it down in a few long gulps.

Most penguins, however, fought the feeding from start to finish, shaking their heads violently and dislodging the fish we had worked so hard to get into them. So we would have to begin the whole process over again until we managed to get four or five fish into each struggling creature. But if the feeding procedure itself started to stress a penguin to the point of exhaustion, we temporarily aborted the process and moved on to the next bird, giving the distraught penguin some time to recover before attempting to capture and feed it again. We spent the remainder of our first day and long into the night repeating this complicated routine, moving from pool to pool feeding hundreds of oil-covered penguins, never stopping once to eat, drink, or even use the bathroom. The thought of stopping for any reason was simply unimaginable when there were 16,000 hungry–albeit uncooperative–mouths to feed.

Finally, at 11 p.m., after sixteen grueling hours of force-feeding penguins, Jay returned to the room and told us it was time to wash our hands and return to our hotel. Covered head to toe with oil, coal dust, fish guts, sweat, and guano, our exhausted team straggled out of the building and climbed into the waiting minibus under the darkened skies of the South African night. As it had been that morning, our group was subdued on the ride back to the hotel as we reflected on the events of the day, the gravity of the situation, and the scope of the work ahead of us. By the time we reached our hotel, the kitchen had been closed for more than an hour; but after seeing the condition of our hungry, bedraggled group, the hotel manager persuaded the chef to stay long enough to prepare a light meal for us. With the horrendous stench of the warehouse emanating from our bodies and our clothing, I imagine he was just grateful we weren't in his restaurant during regular dining hours. The chef and manager graciously served us some pasta and sandwiches, and our team ate without saying much as we absorbed the stark reality of everything we had seen and experienced during our first full day in Cape Town.

Jay and Linda announced that we would gather after our meal for a debriefing meeting to review the rehabilitation strategy and determine which aspect of the animal care we each would be managing. As I glanced at my colleagues around the table, each person looked slightly apprehensive, and my expression undoubtedly mirrored theirs. The sheer size of this rescue operation was hard to grasp, and was far beyond anything we could have imagined before walking into that vast warehouse on that cold July morning. Throughout the day I had heard many people remark that the rescue center felt like a war zone for penguins. The scene inside that building was surreal, and this was indeed the phrase that most accurately described the chaotic atmosphere and shell-shocked demeanor of the birds. We had come to South Africa determined to do everything in our power to help save the oiled penguins. In that moment, however, our years of animal care experience did little to reassure us as we anticipated the tremendous responsibilities in store for each of us, and the implications for the future of the species if we failed.

It was well after midnight by the time we finished eating. We left the restaurant and shuffled down the hallway, following Jay to his hotel room. Propped up against beds and walls, more than a dozen of us sprawled haphazardly on the floor, listening intently as Jay outlined the rehabilitation strategy for the penguins. He and Linda had been observing us throughout the day and now asked for clarification about our individual areas of expertise. Although their team had managed many rescues of oiled wildlife over the years, there were few people at the Salt River Penguin Crisis Centre with previous penguin experience, so our knowledge about husbandry and handling–as well as our experience training volunteers–was essential at this stage. Jay asked for feedback about what we had observed during the day and what changes, if any, we thought should be implemented. We discussed the ideal strategies for rearing abandoned chicks, managing the health of oiled penguins as they waited to be cleaned (a wait that, for many birds, would be at least a month), modifying food preparation procedures, and training the hordes of well-intentioned but completely inexperienced volunteers. After a lengthy conversation regarding the best practices for animal care and management, Jay began reading through his list of names with our respective assignments. Each person waited anxiously to learn what his or her role would be.

Lauren DuBois, supervisor of birds at San Diego's SeaWorld, and Steven Sarro, curator of birds and mammals at the Baltimore Zoo, were charged with the important task of raising hundreds of penguin chicks that had been abandoned in their nests when their oiled parents were removed from the islands. Gayle Sirpenski, animal management specialist at Connecticut's Mystic Aquarium, was given the job of overseeing the food preparation for the penguins; no easy feat, as it required thawing and rationing more than 5 tons of sardines every day. Martin Vince, assistant curator of birds at Riverbanks Zoo in South Carolina, would assist Gayle and help out wherever else he was needed. Alex Waier, supervisor of aviculture at SeaWorld in Orlando, Florida, would be responsible for training a team that would tube-feed every penguin a charcoal substance to counteract the effects of oil ingestion and an electrolyte solution to rehydrate them. Jill Cox, animal care keeper from Utah's Hogle Zoo, was assigned to help Mike Short, a wildlife officer with Queensland Parks and Wildlife Service in Australia. Together, they would be running Room 4, which held about 3,000 penguins. Finally, Heather Urquhart and I, from Boston's New England Aquarium, were told we would be overseeing the care of the 4,500 oiled penguins in Room 2. These assignments from Jay marked the end of our first day as part of the largest penguin rescue ever undertaken. After just one day in South Africa, we were now officially part of IFAW's Oiled Wildlife Rescue Team. There were essentially no other instructions on how to proceed from this point. We were expected to just get in there and do it.

Jay, Linda, and Sarah were now relying on us to draw upon our knowledge and experience to help manage this massive rescue effort. There were fifty-one holding pools in Room 2—*our* room now—each of which held nearly twice as many penguins as we had in our entire colony at the New England Aquarium. Back in Boston, we ran our exhibit with the help of three to four volunteers each day. In our room at Salt River, we would have between 100 and 250 volunteers, working three separate five-hour shifts, to manage every day. The sheer numbers of penguins and volunteers were staggering; but I tried not to think about it, lest I become completely overwhelmed and give up before we even got started.

The only task we had performed all day was force-feeding hundreds upon hundreds of hungry, wild, oil-covered penguins, and suddenly we were part of the rehabilitation management team. The fate of those 16,000 penguins at Salt River now rested partially on our shoulders. It was a heavy weight—against seemingly insurmountable odds—but one that we each bore willingly. Everyone on our team had dedicated much of their lives to working with penguins, from educating the public about their plight to participating in conservation work to increase dwindling populations. Now we were being given the rare opportunity to do something tangible to help save a threatened species, and we were at once awed and humbled.

Back in my hotel room, as I stood under the hot shower washing the stench of the warehouse from my weary body, I marveled at how I had arrived at this moment in time and history. I mentally traced the events and decisions in my life that had brought me to this extraordinary juncture. There was a fleeting moment of panic as I considered how just two of us could possibly manage to run that room with 4,500 oiled penguins in it. Then I climbed into bed, laid my head on the pillow, and, before taking even one full breath, fell into the deepest sleep of my life.

2

Dolphin Dreams and Penguin Pursuits—The Seeds Are Planted

The MOMENT one definitely commits oneself, then Providence moves too. Whatever you think you can do, or believe you can do, BEGIN IT. Boldness has genius, power and magic in it.

—WILLIAM HUTCHINSON MURRAY
(OFTEN MISATTRIBUTED TO GOETHE)

Most people automatically assume, upon learning that I spent the last fifteen years taking care of penguins and teaching people about them, that I've been a crazed and obsessed penguin fanatic since childhood. This simply is not true. In fact, I'll share a dirty little secret with you: I was never particularly into birds. Truth be told, I've always been more drawn to mammals. For as long as I can remember, mammals of all sorts have made me go all soft and mushy: dogs, cats, horses, bears, seals, lions, giraffes–you name it. If it has fur or whiskers, a shiny wet nose, and large, expressive eyes, I'm a goner. And perhaps because I grew up near the ocean, I've always had a particular affinity for marine mammals–especially dolphins. So, to set the record straight, I've actually been a crazed and obsessed *dolphin* fanatic for most of my life. Eventually, though, through a series of amazing chance circumstances, I found myself working with penguins, and it wasn't long before I came to truly love and appreciate these incredibly charismatic and special birds.

I first encountered penguins when I was eight years old during a field trip to the newly built New England Aquarium in Boston. I remember just two things from that visit: the enormous skeleton of a baleen whale hanging from the ceiling and the penguin exhibit, which, at that time, was just a small rectangular display that housed about a dozen penguins. The penguins themselves were smaller and louder than I had expected, but what I recall most vividly was the horrendous stench emanating from their enclosure. Even the solid, thick glass wall between us could not hold back the pungent smell of penguin guano, which seemed to defy the laws of nature by seeping through the glass to reach us where we stood on the other side. Watching the penguins that day, enjoying their boisterous chatter and holding my nose in an attempt to block out the odor, I could never have predicted that twenty-six years later I would spend my days inside of the aquarium's new penguin exhibit caring for some of those very same birds. And I could not have imagined that I would actually develop close personal relationships with them and with dozens of their descendants.

A lover of animals and nature from an early age, I had long been sensitive to the plight of endangered species, but I often felt overwhelmed and powerless in the face of such widespread problems. Though I longed to do something to help save endangered animals, I didn't believe I would ever have the opportunity, let alone the ability, to make a significant difference for any of the world's vanishing creatures. My prevailing childhood dream, however–which I was too intimidated to pursue, imagining how much competition there would be for what I thought was the coolest job on the planet–was of working with dolphins.

My obsession with dolphins started innocently enough at a young and impressionable age. Although many kids are lucky enough to visit a theme park at some point during their childhood, I wonder if there are others besides me for whom it truly became a life-changing experience. During a family vacation when I was five years old, my parents took me to the Miami Seaquarium in Southern Florida. I was already an avid dolphin lover and a faithful devotee of *Flipper,* the television show about two young brothers who enjoyed a special relationship with a wild dolphin whose friendly nature was surpassed only by his remarkable intelligence.

But my obsession with becoming a dolphin trainer and communicating with animals began in earnest that warm spring morning in 1966 as I stood next to the dolphin tank in Florida, watching those sleek and powerful creatures leaping out of the water to seemingly impossible heights. They performed incredible acrobatic feats with apparent ease and, with their long mouths fixed in permanent grins, truly seemed to be enjoying themselves. To this day, I can still picture their trainer standing high above the pool on a platform designed to resemble a ship's bow, a fish dangling from his teeth as one of the dolphins, upon a barely perceptible signal, launched out of the water with startling speed and pulled the fish right out of his mouth with its toothy beak. Incredible! I was mesmerized and stood planted next to that tank watching the dolphins' watery ballet until my parents had to literally drag me away. What captivated me was not only the beauty and grace of the dolphins, but the amazing bond of trust between them and the man standing on that platform. Their ability to communicate with each other without words fascinated me, and in that moment I knew that I, too, had to have that experience someday. But I had no way of knowing that the seed of a powerful dream was being planted that morning.

Although my desire to work with dolphins was a constant background fantasy throughout my childhood and beyond, I always believed that being a dolphin trainer was something that happened to other people. Even if I had thought it possible, I had no idea how to pursue such an unusual career, and my doubts about the reality of making a living doing what I loved prevented me from pursuing this dream. While trying to determine what I wanted to do, I enjoyed stints as a ski bum, a veterinary assistant, and a waitress. Eventually, I became a silversmith and, for eight years, made and sold handcrafted jewelry. In time, I designed a series of endangered species pins depicting various animals threatened with extinction, and donated a portion of the proceeds to organizations working to protect animals and the environment. I saw this as a small way that I could contribute to the protection of animals whose populations were rapidly shrinking. However, it still didn't feel like I was doing enough to make any real difference for endangered species around the globe.

Then, for my thirtieth birthday, my parents gave me an Earthwatch expedition as a present. Earthwatch Institute supports the research of PhDs and field scientists worldwide, and their volunteers spend one to four weeks in the field assisting scientists to collect their data. I chose a dolphin project at Kewalo Basin Marine Mammal Laboratory, an open-air facility overlooking the Pacific Ocean in Hawaii. I spent four glorious weeks there in January 1992, fully immersed in the magical world of dolphins, finally realizing my dream of interacting with these intelligent animals. I learned how to communicate with them using a gestural sign language invented by Dr. Louis Herman, and before long, I could recognize each of the four dolphins–Elele, Hiapo, Akeakamai, and Phoenix–and their individual personalities.

My Earthwatch adventure was truly a life-altering experience, and the springboard for the next phase of my professional life. I reveled in my daily interaction with the dolphins, and realized at the end of my month in Hawaii that there was no avoiding the inevitable anymore. I had to pursue my long-held dream of working with them, no matter what the obstacles might be. I forced myself to ignore the internal voice that told me I was too old to be embarking on this new and seemingly unrealistic venture. And I ignored the naysayers who said there would be too much competition for the job I wanted, and no money to be made. I knew that if I did not at least *attempt* to become a dolphin trainer, I would look back at the end of my life and always regret not having tried. My mantra became: "*Someone* has to have that really cool job of working with dolphins, so it might as well be me!"

I learned that the lab hired several interns throughout the year, but certain areas of academic study were a prerequisite. Undaunted, I found a local school offering a degree that would qualify me to apply for an internship, and the following September, I began classes for a bachelor's degree in veterinary technology at Mount Ida College in Newton, Massachusetts. Part of the requirement for graduation included a year of full-time internships. These were traditionally done at facilities in the Boston area, but I petitioned to do one of my rotations at Kewalo Basin, and much to my delight, my request was approved. In the spring of my sophomore year, my childhood fantasy materialized when I was selected as one of their summer interns.

I spent the summer of 1994 at the lab in Hawaii, deliriously happy and pinching myself daily to ensure I wasn't dreaming. After several weeks of training, I was finally certified as a dolphin trainer. As my internship drew to a close, I cried bitterly at the thought of leaving the dolphins behind; I couldn't imagine anything else could possibly make me feel as fulfilled and euphoric as I did while working with these intelligent and engaging animals, and I planned to pursue a job working with them upon graduation.

That plan changed unexpectedly, however, during my last year of college when I accepted a position as an intern in the New England Aquarium's Penguin Department. From the moment I set foot in their penguin exhibit, the trajectory of my life was altered. Between my first visit to the aquarium in 1969 and my internship there in 1995, an enormous state-of-the-art penguin exhibit had been built, and the colony had grown to fifty birds. This new exhibit took up the entire lower level of the building, covering an area 120 feet long by 70 feet wide. There were five breeding islands in the exhibit, and 135,000 gallons of filtered salt water were pumped in from the harbor. The colony now consisted of thirty-five African penguins and twenty-five Rockhopper penguins–a dramatic-looking species with an ear-splitting vocalization, and yellow and black crest feathers that start above their eyes and droop down the sides of their heads. These crest feathers look like crazy eyebrows, giving them the appearance of being permanently ticked off, and fittingly, this species has a reputation for being quite ornery and argumentative. In fact, on the Rockhopper island in the aquarium's exhibit, if one Rockhopper gets too close to another's space, a heated quarrel inevitably ensues, and within moments, the entire island erupts in high-pitched, excited squawking.

Every day for four months I assisted the staff with feeding the penguins, cleaning their exhibit, and monitoring their health. Three times a day we changed into bulky wetsuits, climbed down a ladder into the exhibit, and waded through the cold, chest-deep water to hand-feed capelin, smelt, herring, sardines, or anchovies to each individual penguin. The number of fish each bird ate was recorded, and notes were made about their general health and behavior. After the morning feeding, the entire exhibit was meticulously cleaned. Every penguin island was scrubbed down using nylon brushes and veterinary disinfectant, and a pool vacuum was used to clean the bottom of the entire exhibit. Feeding the penguin colony took two people about an hour and a half, and cleaning the exhibit took five or six people about three hours. In the winter months, when the water was particularly frigid (about 58°F), my feet always went numb long before we completed the morning cleaning. After climbing out of the exhibit each day, I rushed into the shower and stood under the hot water, waiting for the intense throbbing ache, followed by the inevitable painful tingling that occurred once the blood started to reenter my blanched feet.

At the beginning of my internship I spent most of my time in the "holding room," an area located behind the scenes that temporarily housed penguins taken off exhibit for various reasons, and where chicks were hand-reared. It was here that I helped raise a young African penguin named Sanccob. Sanccob was about seven weeks old when my internship began, and I had the good fortune to help feed and care for him during the period when young penguins form attachments to their caretakers. (Their attachment to us naturally diminished once they reached sexual maturity and started courting other penguins). Sanccob was a very handsome bird, and he seemed to know it. He was quite robust and had a remarkably self-assured attitude, especially for such a young penguin. Most of his fluffy brown chick down had fallen out by the time I met him, having been replaced by sleek, silvery-gray feathers on his back and off-white feathers on his belly–the juvenile coloration for this species. There were still a few small patches of fluffy down clinging to his shoulders, making him look like he was sporting fuzzy epaulettes.

Below his chin was the hint of a darker stripe of feathers, which would become a distinct black band running in a horseshoe shape all the way down the sides of his body when he molted into his adult plumage the following year. And during that same molt, the light-colored spots located just above his eyes would mark the starting point of well-defined white stripes that would loop from the tops of his eyes around his cheeks, and continue down the sides of his stocky, football-shaped body.

Every African penguin has a unique pattern of black spots on its white breast, by which each individual can be identified, much the way every human being can be identified by their fingerprints. Each penguin retains the same spot configuration throughout its entire life, even replacing those few black feathers in the exact same pattern during each yearly molt, when the bird sheds its old feathers and replaces them with new ones. As with all juvenile African penguins, Sanccob's individual spot pattern had already emerged, dotting the off-white feathers on his broad chest and belly. Once his waterproof feathers had fully grown in, Sanccob was returned to the penguin exhibit, where he eventually staked out his own territory on one of the islands. I was pleased that whenever I donned my wetsuit and entered the exhibit, he instantly recognized me and swam over to greet me. He would bump up against my arm, and swim along beside me while I walked through the deep water, maintaining physical contact the whole time. As he escorted me across the exhibit, he puffed up the feathers on his face and cooed to me softly–the penguin equivalent of flirting. It was flattering that he knew who I was and chose to be near me. He was the first penguin I had helped raise, and his response gave me confidence that I hadn't entirely screwed up my first effort as a surrogate penguin mom.

Sanccob was a very sociable bird, and he soon became an ambassador for his species. After I had been hired as a staff member, Sanccob accompanied me on offsite programs to schools, as well as for presentations conducted from inside or next to the penguin exhibit. As with all of our penguins at the New England Aquarium, Sanccob's name had educational significance–he was named after the premier penguin rescue center in South Africa. As one of the main companion birds for our educational presentations, Sanccob grew quite fond of me, and also of Marcia Handin, who had been a volunteer at the aquarium for fifteen years. Marcia had been instrumental in creating our penguin talks, and frequently did these presentations as well. As we endeavored to enlighten the public about serious topics, such as the importance of conservation and the devastating effects of oil spills and overfishing, Sanccob often made it difficult for us to keep a straight face. Much to the amusement of our audiences, he would start flirting shamelessly with us, courting us for all he was worth and attempting to mate with our forearms whenever he saw the slightest opportunity.

A male penguin's mating overture is a rapid beating of its wings against the object of its affection, essentially hugging the sides of the object–which, under normal circumstances, is a female penguin's torso. But during our talks, Sanccob often approached one of our outstretched arms–either in the water or while on dry land–and started slapping his wings frantically against the sides of our forearm, his bill vibrating rapidly back and forth, and his head and neck feathers raised in his excited state. While it was always an amusing display, we did not encourage this behavior. Although we were flattered that he felt so strongly about us, we wanted Sanccob (and the other penguins that sometimes did this) to direct his amorous feelings toward an actual female penguin. So we would gently extract our arm, which usually halted the mating attempt.

On occasion, though, something other than a female penguin or a human forearm would ignite a penguin's desires. And you never knew what it might be. One morning while I was in the penguin exhibit for the morning cleaning, Sanccob came up with a completely new and rather unusual focus for his lusty pursuits. Every day, we vacuumed the bottom of the penguin exhibit to suck up the penguin droppings, along with any objects that were accidentally dropped into the water by our visitors. The pool vacuum has a long, snaking hose about six inches in diameter that floats on the surface of the water, trailing behind the person who is vacuuming. Hearing some frantic splashing and wing-slapping in the water behind me, I turned to find

Sanccob enthusiastically directing his ardor toward this floating hose. He was swimming along atop its entire length, wings beating wildly against the blue plastic, feverishly attempting to mate with it. I burst out laughing, wishing I had a video camera on hand to capture this comical and rather undignified moment in Sanccob's life. My guffawing startled him long enough to stop his frenzied activity and glare at me, with an expression that seemed to say, "Do you mind keeping it down? I'm trying to get lucky over here."

Another African penguin, named Demersus, had a brief but torrid love affair with a black Wellington boot while he was in the holding room for a spell. Whenever I let him and his mate, Ichaboe, out of their enclosure so I could clean it, Demersus made a beeline for the boot. He would start flirting with it, cooing to it softly, and eventually he'd make a full-blown mating attempt, his wings slapping rapidly and loudly against the stiff rubber sides of the boot. Never mind that he was very strongly bonded with Ichaboe. For some reason, he found this tall, shiny boot to be just as desirable and enticing as she was–perhaps even more so. And for her part, Ichaboe acted like a jealous lover whenever she saw Demersus trying to copulate with the boot. She would turn up the volume on her own flirting attempts, doing everything in her power to lure him away from her new rival. She'd do little waddling dances in circles around him, the feathers on her neck and head raised, and her neck curved over gracefully like a swan's. She'd tilt her head and look at him sideways, emitting the low, rumbling growl that is often a prelude to mating. Ichaboe used every tool in the penguin courtship arsenal, trying to get him to turn his attention back to her. Sometimes, in a fit of jealous rage, she attacked the boot, biting and wing-slapping it in an effort to drive it off. In the end, Ichaboe usually won Demersus over, causing him to redirect his romantic endeavors back toward her.

Although the penguins were often a source of laughter and entertainment, there were times when our relationships with them gave rise to anxiety and heartache. Working so closely with these engaging animals every day, we truly came to love each and every one of them, and we often worried about them–even on our days off. When you work with animals and are entirely responsible for their well-being, they are on your mind 24/7. Fortunately, we had a very healthy colony and an excellent veterinary team; yet, like all living creatures, occasionally a penguin fell ill or grew frail with age. It was during those rare but stressful occasions when a health crisis arose that I lost plenty of sleep and chewed my cuticles until they were ragged. It was always distressing when one of the penguins died, whether from illness or old age. As millions of pet owners know, painful decisions regarding end-of-life care for an animal must sometimes be made. At the aquarium, we were occasionally faced with these same tough choices, and we grieved deeply at the loss of any of the animals under our care. Those of us who took care of the aquarium's penguins loved them just as much as we loved our own pets, and we thought of them as part of our extended families. I know I certainly did.

There was another penguin who truly captured my heart during my internship at the aquarium. He was an African penguin named Spheniscus (after *Spheniscus demersus,* the genus and species name for African penguins, which loosely translates to "wedge-shaped plunger"), but he was affectionately called Sphennie by the aquarium staff and volunteers. Along with Sanccob, Sphennie was temporarily residing in the holding room, where he was being treated for a chronic sinus infection that stemmed from a congenital defect in his sinus cavities. A few times each day I had to remove Sphennie from his enclosure and gently pry open his beak to place a medicated gel deep into his choana (a groove in the upper palate), which was red and irritated due to his infection. This was followed by a thorough flushing of his nasal passages with a sterile solution. To do this, I had to insert a syringe with a curved tip into each nostril, infusing 30 milliliters of fluid into each. Most penguins would have protested quite vehemently to this type of handling and medical treatment on a daily basis. But Sphennie always endured the procedure calmly and patiently. He never once tried to bite me, even though the process was surely quite uncomfortable for him.

Not only did Sphennie tolerate this manhandling good-naturedly, he never seemed to hold a grudge against me for it as most penguins might. While penguins will form attachments to some people, they will also form aversions to others–often lashing out at those who have frequently held them for blood draws or other medical procedures. But Sphennie never appeared to resent me for carrying out these treatments. It seemed he intuitively understood that I was trying to help him feel better. There was never any aggressive behavior toward me or evasive maneuvers on his part. Instead, Sphennie regularly "talked" to me–greeting me with a soft guttural *haw* each time I entered the room, and calling out again more urgently whenever I left. After learning to speak "African penguin," I would call back to him, and we would have long conversations throughout the day, hawing back and forth like two penguins smitten with each other. Once he was back in the exhibit, we continued our conversations. Even after I was hired as a staff member, whether I was in the penguin exhibit or just walking past it, whenever Sphennie spotted me, he'd let out a low, throaty *haw*. It didn't matter who was watching or how foolish I might have looked–I always hawed back, happy that the close relationship we had established during my internship remained solid throughout the years.

Years later, when it became clear that Sphennie's chronic and incurable sinus condition had greatly diminished his quality of life, and we could do nothing more to help him, the difficult decision was made to put him to sleep. I was the staff member designated to be with him and hold him while the injection was given that would stop his heart from beating. As the veterinary staff prepared for the procedure, I gently stroked Sphennie and whispered to him what a good bird he had been, telling him how sorry I was that we hadn't been able to help him more. Choking back tears, I kept murmuring endearments to him as the veterinarian inserted the needle. I could hardly bear to watch as Sphennie was given this final injection. After his body went limp and his eyes closed for the very last time, the veterinary staff, sensing my anguish, quietly left the room to give me time alone to say my final goodbyes. I stood hunched over the surgical table for a very long time, cradling Sphennie and bawling like a baby, having just lost one of the first penguins I had really become attached to.

Much to my surprise, during my internship at the aquarium, I found the penguins to be extremely engaging and amusing animals to work with. I had not expected each individual bird to have such a distinct personality or unique temperament, nor had I known that penguins have the ability to recognize their own names. As someone who always had a passion for dolphins and other mammals, I was surprised to discover that penguins were actually quite mammal-like, in both their behavior and their physical robustness. In fact, when visitors asked about their temperaments, I usually found myself comparing them to cats, because if I called a penguin's name and it felt like coming over to me, it did so. If it did not feel like complying, however, it would glance over its shoulder at me (disdainfully, it always seemed) and very deliberately swim away in the opposite direction.

During my four months at the aquarium I realized that there was so much more to learn about these fascinating birds and their distinctive behaviors than I had ever imagined. I fell in love with them and, wanting to remain in the Boston area for a while, decided to pursue a job at the New England Aquarium after graduation. Staff positions didn't open up very often, so I continued to volunteer as a penguin colony associate after my internship ended, hoping to be in the right place at the right time when an opening was announced; a year and a half later, one finally was. After applying, I endured a nerve-racking interview with a group of four department supervisors. Sitting across the conference table from them, I felt as though I was in front of a firing squad. But I was so anxious to work at the aquarium that I would have agreed to lick the guano off the bottom of a penguin breeding cave to get the job, had they asked! When the long-awaited call came offering me a position with the Penguin Department, I eagerly accepted, never suspecting this decision would eventually lead to my participation in the world's largest wildlife rescue, allowing me to fulfill my other childhood dream of doing something tangible to help save a disappearing species.

3

Penguins—Waddling Wonders

In all things of nature there is something of the marvelous.

—ARISTOTLE, 384–322 BC, *PARTS OF ANIMALS*

Penguins are so unique, and so unlike their flying relatives, that many people are not even aware that they are birds. I once encountered a visitor at the aquarium who began arguing with me during one of my presentations, insisting quite vehemently that penguins were actually a cross between a bird and a mammal. I assured him that, as a trained penguin specialist, I was positive that penguins were 100 percent birds. He stubbornly maintained his position while I, as diplomatically as possible, reasserted the established facts about penguins' proper scientific classification. (See Appendix III for a taxonomic rundown.)

The reason for this confusion is easy to understand, and skeptics today share their bewilderment with the early European explorers who first reported sightings of these unusual creatures in the late fifteenth century. When Bartolomeu Dias, and later Vasco da Gama, first saw African penguins (in 1487 and 1497, respectively), they were not sure how to classify these odd-looking animals that were covered with small, tightly packed feathers resembling fur; that brayed like donkeys; and that swam swiftly in the ocean using flattened wings that looked like flippers. Their short wings were covered with minute feathers resembling fish or reptile scales, and, though they could not fly, they collected nesting material and laid eggs the way birds did. Yet their physical robustness and aptitude for swimming and diving in the ocean was more reminiscent of marine mammals. It's no wonder the early explorers were mystified by these strange birds.

Penguins evolved from flying ancestors some 50–60 million years ago during the Late Paleocene and Early Eocene epochs; the oldest known fossil, *Waimanu manneringi,* is some 62 million years old. Penguins closely resemble auks–even filling a similar ecological niche in the opposite hemisphere–and the casual observer often confuses these two groups of birds. Many of us at the aquarium engaged in lively exchanges with visitors who insisted they saw penguins in Alaska or other locations in the Northern Hemisphere, but they likely saw razorbills, murres, puffins, guillemots, or other birds in the auk family. Despite the physical and behavioral similarities, though, auks are not penguins' closest relatives. The closest living relatives of penguins are actually petrels and albatrosses–the latter being marine birds renowned for their phenomenal aptitude for soaring and long-range flight. An albatross can stay aloft for hours without ever flapping its tremendously long wings. A penguin, on the other hand, can flap its stubby little wings as hard as it likes, but it will never get off the ground. How ironic that the flightless penguin should be so closely related to these true masters of the skies.

But why would penguins go through such a radical evolutionary change, anyway? How could it possibly serve a bird to give up the ability to fly? It could only be an advantage if they became better than every other bird at maneuvering in a different medium: the watery womb of the ocean. And that's just what they did. The most likely reason penguins evolved to leave the heavens and the freedom of flight behind was so they could swim faster and dive deeper than all other birds, thereby catching prey that their avian counterparts could not reach. The driving force behind any animal's physical and behavioral transformation over time is to fill an ecological niche, to enable it to survive and thrive in a very specific habitat where few of its competitors can, thus ensuring a source of food and shelter. While penguins have to compete with some large fish and marine mammals for fish, squid, and krill, they have few avian competitors for their main prey items.

But to give up flight, and become aquatic champions instead, penguins' bodies underwent many drastic changes over the millennia. Most birds have hollow, air-filled bones, making them lightweight for flight. In fact, the combined weight of the feathers covering a bird that flies is more than the combined weight of its bones. Penguins, on the other hand, have dense, marrow-filled bones, like those of mammals. This provides extra ballast, enabling them to dive using minimal effort; and the lack of air inside their bones makes it easier for them to stay underwater without having to fight the effects of buoyancy. If penguins had air-filled bones, they would be forced to the surface like a cork the moment they stopped flapping their wings, the way puffins and loons are. Penguins' main prey items are generally found at depths between 35 and 300 feet, so their solid bones are an excellent energy-saving modification for the deep dives they must make to catch their food.

Penguins had to make several other adaptations in order to survive and thrive in the oceans. Water is 800 times denser than air, which is why it feels like moving through molasses when you try to run or walk through it. Evolution helped penguins overcome the impediment of this increased friction in several different ways. To reduce drag in the water, their bodies developed a streamlined football shape, and both their wings and their feathers became much shorter and stiffer than those of their flighted relatives. Their flat wings, which are slightly tapered from front to back like an airplane wing, are powered by massive chest muscles; and where flighted birds only get forward momentum on the down stroke of their wings, penguins move forward on both the down stroke *and* the up stroke by changing the pitch of their wings. Like all birds, they move by flapping their wings up and down; although their method of locomotion is officially labeled swimming, penguins use their strong wings to essentially fly through the water.

Whereas most birds' legs emanate from a point near the middle of their abdomen, penguins' legs eventually migrated to the rear of their bodies. This change in position also contributed to their streamlined form; instead of protruding from the middle of their bellies, where they would disrupt the flow of water, their short legs trail behind them as they swim. And, though penguins have webbed feet, theirs are not used for propulsion as with other aquatic birds, so you will never see them paddling like a duck. Instead, their feet are used solely for steering–and occasionally for reducing speed. As it swims, a penguin lets both feet trail straight out behind it, with the bottom surface of the feet facing up toward the sky. When it wants to change direction, it dips down one foot, pointing its toes toward the inky depths of the ocean. Even the way this seabird turns is counterintuitive. Instead of dipping down its right foot to turn to the right, as one might imagine, it dips down the left foot to turn to the right, and the right foot to turn to the left. And if the penguin needs to slow down quickly, it points the toes on both feet down together at the same time, effectively putting on the brakes.

Our visitors at the aquarium were often amazed at the speed with which penguins could change direction. An informal activity that many visitors (as well as staff) enjoyed playing with the birds was something we called "the shadow game." When someone waved an arm over the penguin exhibit, the lights from above cast the shadow from their arm onto the floor of the pool. Many of our penguins enjoyed chasing these shadows, making repeated small, rapid turns as they pecked at the shadow on the floor, showcasing their agility. Oftentimes, when the person stopped playing, their new penguin friend would swim below them in tight circles, looking up expectantly and waiting for them to continue. And if the penguin was really into the game and the person still didn't respond, it would *haw* loudly at them as if to say, "Come on! Let's play some more."

The trade-off for having a well-placed rudder, however, is an upright, humanlike posture and an ungainly, waddling gait while on land. Despite their awkwardness, penguins can run surprisingly fast, and, since 75 percent of their time is spent at sea, the compromise seems to have served them well enough. While they're not very graceful on terra firma, I've often been surprised at the speed with which these birds can move when motivated. On several occasions, I've witnessed penguins coming ashore at night and rushing to enter the ocean in the morning. These ocean entrances and exits are extremely hazardous moments in their daily lives, as there are often avian and mammalian predators lurking nearby, waiting to pick them off as they head off for, and return from, their fishing expeditions. Because there is safety in numbers, the hypervigilant penguins will wait to make their move, gathering in large groups before running across the beach as fast as their little legs can carry them. They often get moving so quickly that the momentum causes them to fall forward onto their chests and faces; then they start paddling wildly at the sand with their wings and feet, using every means possible to get out of the open and into the relative safety of the ocean. Tripping and stumbling over their own feet and wings as they hustle across the beach, with clouds of sand and dust flying behind them, they look like cartoon characters trying to outrun a dreaded adversary.

At some point during the evolutionary process penguins developed another critical adaptation for living in the ocean–one that they share with sea turtles and other marine birds. Salt glands, located under the skin above their eyes, process and excrete the excess salt in their systems, allowing them to drink salty ocean water to their heart's content without ever becoming dehydrated. These comma-shaped glands essentially act as a second set of kidneys, desalinating the seawater they drink by filtering the salt from their blood. Penguins get the hydration they need from the food they eat and from the seawater they drink, and, during periods of fasting, from metabolizing their fat reserves. So, while most animals must have fresh water to drink, penguins have the remarkable ability to survive without ever having access to fresh water.

Finally, to protect them from the frigid waters they spend most of their time in (and from the harsh rains, winds, and blizzards they must endure during breeding and molting on shore), penguins had to radically change their feather structure. Penguins have more feathers than any other bird on earth–on average, about seventy feathers per square inch. While penguins living in colder regions have even more feathers than those in warmer regions, they all have small feathers that are stiff, narrow, and slightly curved, with downy filaments at the base. Penguins can raise and lower their feathers similar to the way the hair on our arms stands straight up when we get goose bumps, though they can do it voluntarily. To warm itself, a penguin raises all of its feathers and traps a layer of air in the downy base; then it lowers its feathers again, with the pointed tips laid over each other like shingles on a roof. Like all birds, penguins are warm-blooded animals, so their body heat warms this layer of air that is locked in the downy filaments next to their skin. To cool off, the penguin merely lifts its feathers again to release the warmed air. In cold ocean waters, this diving bird needs to stay warm, but as the penguin submerges, some of the trapped air is forced out by the water pressure squeezing against its body. To compensate for losing this warm air barrier, the penguin is protected by a thick layer of fat, which helps it retain heat. This dense layer also helps to cushion the penguin as it gets smashed against the rocks while attempting to come ashore through breaking surf.

It is these birds' tightly packed feathers that allow them to survive and thrive in some of the harshest climates known to man. Their feathers are truly critical to their ability to survive, which is why getting caught in oil spills is so deadly. To stay dry and to ensure that their feathers remain in good condition, penguins spend an inordinate amount of time preening, using their beaks to neatly arrange and oil their feathers. Penguins have a special gland, called a *uropygial gland,* near the base of their stubby tails (on the topside of their rumps), from which they get a light waxy substance that has the consistency of baby oil. Twisting their long necks around to reach their tails, penguins rub their beaks across this gland to collect some of the oil, then use their beaks to spread it across all of their feathers.

Contrary to popular belief, this uropygial oil does not make penguins' feathers waterproof. It actually conditions and softens the feathers, helping them to last longer. The oil provides a minimal degree of waterproofing, but is limited in its ability to do so. As everyone knows, oil and water don't mix–and droplets of water do bead up and roll off penguins' feathers; but their uropygial oil is only partially responsible for that happening. What really keeps water from seeping through their feathers to reach the skin are microscopic barbules on the feathers that lock together like Velcro, forming an impenetrable barrier. But petroleum products–be it heavy crude oil or lighter diesel–destroy the barbules' ability to fasten to each other. The feathers then clump and separate, so that large gaps form across the penguins' coat, leaving them vulnerable to hypothermia, starvation, and drowning.

Preening is truly an activity that is critical to the penguins' survival, yet even with their long, flexible necks, it's difficult to reach certain areas on their own bodies. While they use their beaks to preen their feathers, they can't reach their own necks or heads this way. So, to condition these feathers, a penguin first coats its beak with oil, then transfers the oil from its beak onto the front edges of each wing. Then it raises each wing up high and, twisting its neck, rubs the side and top of its head across the leading edges, thereby oiling the head and neck feathers. When a penguin has a mate, however, that mate can help to preen these hard-to-reach areas. If you are observing a mated pair of penguins, it's common to see them nibbling each other around the head and neck, usually at the same time. They devote a lot of time to this pair-bonding activity, which, when done simultaneously, is called *allopreening* or *mutual preening*. Like their wild counterparts, our penguins at the aquarium spent much of their time preening and mutual preening; and sometimes, even we were the recipients. Sanccob and Sphennie, and other birds who were very comfortable being handled, would very gently nibble the skin on our arms if we scratched the backs of their heads or necks. If this allopreening went on for a while, we ended up with rows of tiny red marks, like pinpricks, on our forearms. It was undoubtedly just an instinctive response, but it seemed like we were getting little penguin kisses.

Even with careful preening and conditioning, over time, penguins' feathers get worn out; they become degraded from the harsh rays of the sun, and clumps of feathers may be lost in fights or in narrow escapes from predators such as seals or sharks. Once the feathers are no longer watertight, a penguin's ability to survive can be fatally compromised. But nature has provided a unique solution to this problem. Once a year, for the duration of their lives, penguins lose and replace every single feather on their bodies in a process called *molting*. While all birds molt, most species replace just a few feathers at at a time, as needed. But because penguins require a complete coat of feathers to remain waterproof, they must replace all of their feathers at once in a more radical process, known as a *catastrophic molt*. During a penguin's molt, which takes two to four weeks, every last feather–including the tiny, scalelike feathers on their wings–falls out and new feathers grow in. The new feathers actually form under the skin prior to the molt, and push the old feathers out as they grow in behind them–much like our adult teeth push out our baby teeth as they grow in.

This molt takes a tremendous amount of energy, and, as with all living things, food is required to produce energy. But once the penguins have lost all of their feathers and are essentially bald, they would immediately freeze if they attempted to enter the ocean to forage for food. Instead, they must remain on land until their new feathers have completely grown in. To survive this lengthy fast, penguins must carry out a protracted feeding orgy in preparation for their molt. For four to six weeks prior to molting, penguins spend most of their time hunting, their voracious appetites spurred on by the changing hormones that precipitate the annual molt.

Our penguins at the aquarium exhibited this same pre-molt hunger, and these ravenous birds would pester us relentlessly during feedings. They would be in a virtual frenzy, pushing other penguins out of the way and knocking them off the island in their fervor to reach us and the fish in our buckets. If we didn't feed them quickly enough, they would hop into the water and circle us, nipping at our arms and sides until we paid attention to them. During this pre-molt phase, penguins gorge themselves, increasing their body mass by some 30 percent–even their flat wings swell to more than three or four times their normal thickness. Once the newly formed feathers are ready to push through the skin to replace the old feathers, penguins eat their last meal and haul out onto shore, where they stand–barely moving and looking a wreck–for the next few weeks. During their molt, penguins become very cranky–I can only assume because they are extremely uncomfortable and because their hormones are raging. They don't want to eat, don't want to move, don't want to be touched, and will lash out and bite at anyone or anything within reach. I've often jokingly compared it to PMS times twenty.

As the catastrophic molt progresses, clumps of feathers start falling off their bodies, leaving large expanses of bare skin exposed. Molting penguins look so ratty, disheveled, and miserable that visitors at the aquarium often asked if our birds were sick, prompting another lesson on this unusual phenomenon. The physiological stress of molting is extreme, and it is during this time that penguins are most susceptible to illness, hypothermia, and starvation. They no longer have their usual protection against the elements, and in the wild they must suffer through drenching rains, driving snows, and freezing winds half naked and shivering. If they do not succumb to the cold, starve, or fall ill, they will come out the other side of this annual marathon with a shiny new coat of feathers. Once all their feathers have grown back in, the penguins can finally return to sea, where they will spend the next several weeks hunting and eating to regain the weight they lost during their molt. Penguins survive their enforced yearly fasts by metabolizing their fat stores during their molt; most will lose up to half their body mass during this time. If a penguin does not catch enough fish and store enough fat prior to molting–whether due to a scarcity of fish or because of decreased hunting ability as a result of age, injury, or illness–it may not survive its molt.

The feathers covering a penguin's wings are quite different from the rest of the feathers on its body, and even these get replaced during each annual molt. These tiny, rigid feathers resemble overlapping fish scales–there is no downy base to these extremely short feathers, making the wings very flat and smooth and impervious to water. Much like competitive swimmers who shave their bodies to increase their hydrodynamic efficiency, the sleek, unbroken surface of a penguin's body and wing feathers provides yet another means of streamlining. Even the bones inside the penguin's wings differ from those in other birds; in addition to being solid, they are completely flattened, which helps to make effective weapons of their wings. When fighting or trying to defend themselves, penguins rapidly paddle their adversaries with their wings (a behavior called "wing-slapping"). Their wings are surprisingly hard. Powered by strong chest muscles, the force with which the penguins hit is considerable and rather startling. I myself have been on the receiving end of many a wing slap, and have been left with raised welts and nasty bruises from each of these encounters. The speed with which they hit is also astonishing: they can get in several solid whacks before you even have time to pull your hand away. Their flattened wing bones, along with other anatomical and behavioral modifications made over millions of years of evolution, have left penguins superbly adapted to life at sea. Swimming faster and diving deeper than any other birds, penguins zip through the water like speeding torpedoes; and while they generally cruise along at approximately 5 miles per hour, these seabirds are capable of reaching speeds up to 15 miles per hour or more in short bursts.

Penguins have thrived in the oceans of the Southern Hemisphere for millions of years, but they have encountered many threats to their survival since their discovery by man. For some species, the consequences have been worse than for others. For penguins living in warmer climates, such as the African and Galápagos penguins, the impact has been greater. Although our negative influence on the environment has now reached the remote Antarctic as well, affecting the penguins living and breeding in this once pristine location. Sadly, for all penguin species, contact with humans has proved to be a losing proposition.

4

African Penguins—A Species on the Brink

We are at a critical juncture, with the African penguin population in apparent free fall.

—PETER RYAN, PERCY FITZPATRICK INSTITUTE OF AFRICAN ORNITHOLOGY, UNIVERSITY OF CAPE TOWN

While there are currently eighteen recognized species of penguins scattered throughout the Southern Hemisphere, African penguins are the only penguins found on the African continent. These birds have long led a hardscrabble existence on a number of small islands off the coasts of South Africa and Namibia, and they have been dealt several harsh blows since being discovered by European explorers. The earliest threats were from decades of unregulated harvesting of their eggs and of guano. For thousands of years, African penguins had constructed nesting burrows out of dried guano, using their webbed feet and long, curved toenails to dig into the deep layers of seabird droppings that covered their islands. Having built up slowly over many centuries, the guano (from the penguins themselves, as well as from other marine birds such as gulls, cormorants, and shags) had a firm, claylike consistency, so when the penguins hollowed it out to form their breeding caves, the structures held their shape and didn't collapse. And because it was porous, when there were heavy rains, instead of running in rivers through the colony, flooding their nests and drowning their chicks, the water was evenly absorbed by the deep guano cap. On some of the penguins' breeding islands, the birds nested on steep slopes, and these caves prevented eggs and chicks from rolling out of their nests and tumbling down the hillside. More critically, these cool, shaded burrows provided shelter from the scorching sun, as well as protection from terrestrial and avian predators.

But in the early 1840s, enterprising Europeans and Americans discovered these ancient seabird deposits–in some cases more than 80 feet deep–and, realizing that the nitrate- and phosphorus-rich excrement could be harvested and made into fertilizer, scraped the islands down to the bare rock. The guano was so valuable that it was called "white gold." The harbors and bays of the penguins' nesting islands were clogged with the ships of guano collectors looking to strike it rich. The islands of Namibia–some with fanciful names such as Roast Beef and Plum Pudding–were particularly hard hit; at one time, Ichaboe Island in Namibia was surrounded by more than 400 ships. In a three-year period in the mid-1840s, at this tiny 15-acre island alone, a 60-foot-deep layer of the reeking stuff, weighing 800,000 tons, was removed. To supplement their diet, the 6,000 men working there also ate the penguins and their eggs, further reducing their population. Having stripped Ichaboe Island of its thick guano cover, the ships' crews proceeded to make their way through the rest of the African penguins' breeding islands in Namibia and South Africa, systematically removing the seabird droppings until nothing was left. In the space of just a few years, the guano that had been accumulating on these rocky offshore islands for thousands of centuries was completely eliminated.

Leaving no island untouched, this massive harvest destroyed the penguins' nesting habitat, forcing them to incubate their eggs and raise their chicks out in the open, where they were often fatally exposed to the elements and to predators. The seabird deposits have never built back up to any significant level since the great guano craze, and the

African penguin population continues to suffer from these activities, driven solely by monetary gain, that took place more than a century ago. Despite laws passed in 1967 prohibiting the collection of guano, the harvesting continued unchecked on several islands until 1991, and continues even today in parts of Namibia. The following quote is taken from the website of Guano Green Fertilizers, a company based in South Africa: "In July 2003 Guano Green Fertilizers Ltd. secured a 70% interest in a joint venture with Namibian Guano Islands. The rights to harvest guano on the islands of Ichaboe and Mercury off the Namibian coast have been secured." While there may be other guano-harvesting companies operating in the region, this one company alone has been granted permission to remove 4 million pounds of guano per year from two of the African penguins' main breeding islands in Namibia. Separated geographically from their South African counterparts by approximately 1,000 miles, the Namibian segment of the population is highly endangered, so this practice could have tragic consequences for the penguins living and breeding there.

Not long after the guano harvest in the nineteenth century, a devastating egg harvest began on the islands. Combined with the long-term negative impact the removal of guano had on their reproductive success, this egg harvest had a catastrophic effect on the African penguins, ultimately causing their population to crash. Starting shortly before the turn of the twentieth century and continuing unchecked until 1969 (when a ban was finally enacted), millions of eggs were taken from the African penguins' breeding colonies. Collected commercially for human consumption, the eggs were initially used as an inexpensive source of protein for the poor; but eventually–as stocks decreased and demand rose–the penguins' eggs became a luxury food item. As a result, less than half of the eggs that were laid were left behind to hatch during dozens of successive breeding seasons. From 1900 through 1930, nearly 500,000 eggs were taken each year from Dassen Island. In the space of thirty years, 13 million eggs were collected from this one island alone. This represents 13 million penguin chicks that never had a chance at life, and that's from just one of their two dozen breeding islands.

At the turn of the twentieth century, there were approximately 3 million African penguins on earth, and Dassen Island was home to half of them. Today, primarily due to egg and guano harvesting–and in more recent years, overfishing–there are just 5,000 pairs breeding there. Photographs taken on Dassen Island in 1914 show penguins covering every square foot of the island. Less than twenty years later, photos taken from the same vantage point paint a much bleaker picture: these images are eerily devoid of penguins. Large open expanses of flat earth with shallow borrows scratched into the surface can be seen, with only a few scattered penguins dotting the barren landscape.

The guano and egg trades alone might have ultimately doomed the African penguins. But adding more fuel to an already blazing fire was the overfishing of anchovies and–more critically–pilchards (also called sardines), the preferred prey of these seabirds. During the 1960s, after years of unregulated commercial fishing of pilchards along the west coast of South Africa, the population of these once abundant fish crashed. Despite the introduction of laws restricting the catch, the pilchard population has never fully recovered in this part of the world. While not reaching back as far in time as the destructive practice of collecting their guano and eggs, the more recent depletion of the African penguins' primary food source has nonetheless significantly contributed to the species' decline. Not only have adults starved; parents haven't had enough fish to feed their chicks, leading to a total breeding failure on some islands. On Possession Island in Namibia, the population of African penguins fell from 23,000 breeding pairs in 1956 to just 500 pairs in 1987 due to this food shortage. After the local pilchard and anchovy stocks collapsed, the penguins were forced to search out new and less abundant prey items, such as gobe, red hake, and squid, but it was already too late. The population of African penguins in both South Africa and Namibia continued to plummet.

Another tragic side effect of the fishing industry has been the entrapment of penguins in fishing nets. Attracted by the same schools of fish that the fishermen are pursuing, the penguins often get ensnared in their large nets, and as air-breathing birds that can only hold their breath for a few minutes, these penguins drown long before the crew has hauled up their catch. Once the nets are raised and the dead penguins are discovered, they are removed from the nets and their lifeless bodies are tossed back overboard into the sea. Fishing net entanglement–in trawl nets, gill nets, and purse-seine nets–is a long-standing problem that involves both commercial fishing vessels and small-scale, local fishing boats. Researchers have encouraged fishermen in various parts of the world to report this unintended bycatch; but even without fear of retribution or fines, reporting of these accidental penguin drownings is not consistent, and it is not known how many penguins die each year in this way. Even the monofilament fishing lines left behind by individuals out for a day of recreational fishing present a major hazard. These invisible lines have entangled and killed many penguins in South Africa and throughout the rest of the Southern Hemisphere. I've seen many heartbreaking images of penguins, both alive and dead, bound and wrapped tightly in yards of discarded fishing line.

Longline fishing is yet another lethal threat that penguins must contend with. A more recent scourge of the seas, this method of fishing kills approximately 100,000 albatross and 200,000 other marine birds every year (in addition to millions of fish, marine mammals, sea turtles, and other unintended victims); and it is an apparent menace for penguins as well. A single fishing vessel can trail an 80-mile longline behind it, containing as many as 20,000 baited hooks. A hungry seabird, spotting the bait on or near the surface, swallows it along with the hook, then is dragged underwater and drowned as the line sinks. Fishing vessels bait approximately 3 *billion* longline hooks each year, and albatross are being killed by this fishing method at the alarming rate of one every five minutes. In South African and Namibian waters alone, approximately 33,000 albatross are drowned every year after getting snagged by longline hooks (along with 4,000 endangered sea turtles and 7 million sharks); the number of African penguins that are inadvertently caught is not known. Fortunately, due to recent conservation efforts spearheaded by BirdLife International and World Wildlife Fund for Nature–South Africa (WWF-SA), the chilling number of albatross killed every year by longlines is starting to decrease. By educating the fishermen about new techniques that reduce seabird bycatch, and by establishing and enforcing limits on the number of birds that can be accidentally caught, the lives of thousands of albatross are now being saved. Hopefully, penguins and other seabirds will be saved along with them.

Egg collection, guano harvesting, overfishing, and entrapment in fishing gear: all of these threats, as well as habitat encroachment, environmental pollution, and introduced predators that devour eggs, chicks, and adults, have imperiled the African penguins. And, like many aquatic animals, they now suffer from plastic ingestion. All this has led to their being listed by the International Union for Conservation of Nature (IUCN) as a species "Vulnerable to Extinction." The population crashes on Dassen Island and Possession Island are indicative of what has happened on almost all of the African penguins' breeding islands during the last century, bringing into question the future of this species. The world population of African penguins has plummeted by 95 percent in the last fifty years; and the most recent census numbers reflect an even more rapid rate of decline. Between 2000 and late 2009, their population shrank from 50,000 pairs to a mere 25,000 pairs, a decrease of 50 percent in just nine years.

The situation is so dire that penguin researchers have discussed making appeals to have the African penguin declared an endangered species. They have theorized that most of these birds have succumbed to starvation due to overfishing and the displacement of their prey items. This food displacement is brought about by the shifting of the cold ocean currents that the fish they seek thrive and travel in; this shifting of currents, in turn, is brought about by global climate change. There is a normal cyclical change in the movement of sardines and anchovies in this part of the world–but in conjunction with other survival threats, the recent displacement of this primary food source has proved catastrophic. Food scarcity, added to the many other challenges the penguins have faced historically, may be the final insult that ultimately dooms these birds to extinction. Sadly, African penguins have never recovered from the various modes of exploitation inflicted upon them by humans in the last 150 years, and their numbers continue to decrease at a rate that may soon prove to be unsustainable. If more is not done to protect them, African penguins could be wiped out in the wild as early as 2020.

There is one other risk these vulnerable seabirds face on a daily basis. Because they live in the midst of a shipping superhighway for tankers and cargo ships, oil contamination is a perpetual threat. Every day, dozens of ships pass directly through their breeding and foraging grounds, polluting the waters the penguins are entirely reliant upon. Navigating their way across the southern oceans, these enormous ships, heavily laden with ore, oil, and other cargo, are at particular risk in the waters off the South African coast–an area legendary for its rogue waves, powerful currents, and violent winter storms. Even those ships that survive the stormy seas around the tip of South Africa do not leave the region untouched by their presence. A common practice is the illegal cleaning of bilges and holding tanks at sea. After unloading their cargo, the crews of ships and tankers fill the holding tanks with seawater; this serves as ballast, giving the ship more stability and making the next leg of the voyage safer and more comfortable. After arriving at the next port to load their ship with cargo, this ballast water is supposed to be pumped out into holding containers at the dock, along with the bilge water that has collected in the bottom of the ship. However, this costs money and takes time, so to avoid the expense and inconvenience, many crews simply dump the ballast and bilge water out at sea prior to entering their next port.

Not only is this practice illegal; it pollutes the marine environment because the water has inevitably become contaminated with fuel and oil. To avoid detection, ships usually dump this water at night some distance from shore, where it's harder for authorities to patrol, and more difficult to spot smaller amounts of oil on the ocean's surface. While not leaving behind the same amount of petroleum that would escape from a ship during a major spill, the results can be catastrophic for those animals that come into contact with these drifting patches of oil and fuel. Because the amounts of oil left behind are relatively small and there is no reported spill, authorities usually never know about these illegal purges of polluted water. As a result, wildlife rescuers are never alerted, and the oil-contaminated animals are never saved. Even a small amount of oil on feathers or fur can be fatal. Whether it's a penguin, an albatross, or a fur seal, any marine animal that's been covered with oil will suffer a protracted death at sea or on shore unless it is rescued and rehabilitated.

Every week, somewhere on the world's vast oceans, two commercial vessels meet an untimely end and sink, fouling the environment. Some are intentionally scuttled for insurance payouts, while other ships are so old or poorly maintained that their hulls have become compromised due to rust and metal fatigue (once a crack or hole develops in the weakened metal structure, it's not long before the hull suffers catastrophic failure). Whether the result of storms, collisions, negligence, or even a deliberate act, every time a vessel sinks or runs aground, it poses a threat to the animals living in that ecosystem. A cargo ship's fuel oil will leak into the surrounding waters as the ship goes down, and an oil tanker carrying millions of barrels of oil can cause a devastatingly large oil spill as it sinks. Even years after sinking, these ships littering the ocean floor continue to pose a threat to the environment. It is estimated that more than 8,500 commercial and naval vessels went down during World War II alone, carrying up to 20 million tons of oil and fuel with them to their watery graves. As time passes and corrosion takes its toll, this oil will be discharged into the world's oceans. These sunken ships are ticking time bombs, and it is impossible to predict when they will release their deadly caches of oil.

Though all commercial marine vessels must be registered, insured, and inspected, regulations–and enforcement of those regulations–vary from country to country. Many shipowners, to save money on taxes and insurance, and to avoid penalties, register their vessels in countries where the fees are lower, the regulations more lax, and the inspections substandard–if even existent. These are called "open registries" or "flags of convenience," and it's very convenient for unscrupulous or greedy shipowners to register their vessels in some of these countries. Ships flying under certain flags may be poorly constructed and poorly maintained; their crews may be improperly trained, overworked, and underpaid, forced to work in unsafe conditions, leading to even more accidents. Because each country has its own set of laws and safety standards, by registering their vessel under certain flags of convenience, owners can legally skirt the laws designed to make shipping safer for both the crew and the environment. While not all open registries are disreputable, there are some where the standards are laughable and the laws nonexistent. One shocking example was the Cambodian Shipping Corporation (in operation from 1994 to 2002), whose questionable activities and unsafe practices ultimately caused the registry to be shut down. Many countries have open registries that will certify anyone who wants to pay their fees; the Cambodian Shipping Corporation was the only company where shipowners could register their vessels *online* and receive their official papers in twenty-four hours, without the ship ever being seen by officials.

In other efforts to save money, many owners fail to maintain their ships properly, and, in some cases, corrupt inspectors are paid off to ignore glaring structural or mechanical problems and conspicuous breeches in safety protocols. More than one vessel has foundered within weeks of having passed an inspection and, when examined after sinking or running aground, preexisting hull damage has been discovered. In fact, just nine months prior to sinking in Cape Town, an inspection of the *Treasure* revealed twenty deficiencies–yet the ship was not detained. Today, about 4,000 oil tankers transport petroleum products across the globe; with an expected life span of twenty to thirty years, more than half of these ships are now over twenty years old. It's only a matter of time before more vessels in this aging fleet break apart and sink, spilling their enormous stores of oil as they descend to the ocean floor. Despite the potential danger to both crews and the environment, these antiquated tankers are still allowed to make their risky ocean crossings.

For the African penguin, the *Treasure* oil spill was not the first time the future survival of the species had been jeopardized by oiling.

These birds have been struggling to survive spills that have polluted their home waters for nearly a century. The first documented sighting of an oiled African penguin was in 1921; the first recorded major spill occurred with the sinking of the *Esso Wheeling* in 1948. It is not known how much oil entered the environment or precisely how many birds were harmed in this spill, but all of the oiled birds—numbering somewhere in the thousands—died. It is estimated that one third of the penguins living and breeding on Dyer Island perished in this disaster. And during an oil spill of unknown origins in 1952, in which more than 1,200 oiled penguins came ashore on Robben Island, seabird ornithologist Bob Rand made the following observation: "Soiled penguins died on the beaches or lingered on the islands to perish of hunger. Where nesting birds were affected, chicks also died. No matter how small the contamination, the birds refused to take to the water." This sobering description of the aftermath of the oil spill, and the resigned behavior of the penguins, paints a vivid and tragic picture of the impact of oiling on birds. Although penguins are hardy and resilient seabirds, fine-tuned through evolution to withstand the rigors of life at sea, all of their adaptations are useless against just a few drops of deadly oil penetrating their feathers.

But it was the grounding of the *Esso Essen* on April 29, 1968, that inspired one remarkable woman to establish the very first rescue center for penguins in South Africa. Although legislation at the time required ships to be at least 12 miles offshore, the *Esso Essen* was just 3.5 miles offshore when it ran aground, and the 4,000 tons of oil that spilled from its tanks oiled 1,700 African penguins, as well as 1,300 other seabirds. At the time of this oil spill, no formal wildlife rescue center existed. People tried valiantly to save the oil-soaked birds, but there were few established protocols for cleaning and caring for penguins or other wildlife, and almost every bird that came into contact with the oil died.

Althea Westphal, a thirty-seven-year-old British woman living in South Africa, had been involved in the failed efforts to save the oiled birds. Horrified and deeply disturbed by this environmental disaster, Althea was spurred into action. She recognized that, to be successful in saving oiled and injured wildlife, an established rehabilitation center with trained staff using well-researched protocols was needed. People had to be fully prepared to respond in an organized and capable manner to future oiling events, instead of just reacting in a spontaneous and largely ineffective way each time one occurred. And so she built a small rescue and rehabilitation center near Cape Town, which she called SANCCOB. The center was officially constituted in November 1968, just seven months after the grounding of the *Esso Essen.* The following story about Althea, and the establishment of SANCCOB, is taken from SANCCOB's website:

> *From the Westphal's house in Claremont, the birth of SANCCOB occurred in the late 1960s. The* Esso Essen *spill was the first of the major recognized spills and Althea set up a temporary station at her home where she began rehabilitating 60 badly oiled penguins. In those days, the birds were scrubbed with Sunlight soap and fed long strips of hake which had been dipped in fish or sunflower seed oil. The birds were given a 50/50 chance of survival. Althea's birds were washed in her bathroom, three at a time, and then rinsed with a hose. The first swimming pool was a wooden trailer in Althea's garden, after which she obtained a huge stainless steel dye vat. Two or three times a week the birds were driven to Blaauwberg in Althea's station wagon, marched down the beach to the tidal pool and allowed to swim for an hour. The first flipper rings were colored bias binding, and then dymotape and finally G-rings which were supplied by the Percy FitzPatrick Institute (PFPI) of UCT [University of Cape Town]. During this time, Althea carried out extensive research on the "Jackass penguin" to help her understand its lifestyle and dietary requirements. Early in 1968 Althea started enquiries into establishing a rescue operation, and eventually she persuaded Dr. Roy Siegfried of PFPI to help her launch SANCCOB–a task they believed would cost about R150 000 [about $18,000]. Eventually a group of concerned individuals rallied together, including members of PFPI and the SA Army, and created an informal SANCCOB.*
>
> *To obtain official recognition for the African Penguin, proof that the species was declining had to be provided. This proof was supplied with*

photographic evidence of the islands from 1914 and the 1930s. This proof gained Althea a permit to operate by the Department of Guano Islands, and a grant of R10 000 [about $1,200] from the South Africa Wildlife Foundation (now the WWF-SA) for a three year Population Dynamics Study on Dassen Island. SANCCOB achieved its first milestone in December of 1969 at a conference in the Kruger National Park when the collection of penguin eggs on the islands was banned!

Born from one woman's vision, drive, and commitment, SANCCOB has become a highly respected world leader in the rehabilitation of oiled and injured penguins. Without Althea's determination to rescue and heal ailing seabirds, countless animals would have endured horrific deaths from oiling and other injuries caused by human activities. Since 1968, SANCCOB's dedicated staff and volunteers have saved the lives of tens of thousands of penguins and other marine birds. Not only have thousands of penguins been directly saved; generations of offspring from those rescued penguins have been spared as well. In 2002, Peter Ryan, a researcher with the Percy FitzPatrick Institute of African Ornithology, stated: "Due to SANCCOB's efforts, the population of African penguins is 19% higher today than it would otherwise have been." This translates to roughly 33,000 more penguins than there would have been without the existence of SANCCOB. Were it not for Althea Westphal's compassionate heart and practical foresight, this vulnerable species would be one step closer to extinction.

Over the years, several large oil spills have threatened the African penguins. Starting with the *Esso Wheeling* spill in 1948, fifteen spills have oiled significant numbers of African penguins; twelve of these involved more than 1,000 penguins each. There was a marked increase in oil spills between 1967 and 1975 due to the closing of the Suez Canal. Approximately 650 oil tankers that normally would have taken this shortcut between the Mediterranean Sea and the Red Sea each month were forced to travel the long and perilous route around the tip of South Africa and the Cape of Good Hope—appropriately dubbed the "Cape of Storms." Many of these ships were quite old, and not designed for the rough seas of the southern oceans, resulting in an increased number of spills in these waters.

In addition to the penguins affected by these major oil spills, approximately 1,000 inexplicably oiled penguins are cared for by SANCCOB every year. These penguins have been caught in smaller, unreported "mystery" spills, or have swum through the oil-contaminated ballast and bilgewater illegally dumped into the ocean by passing ships. Between 1966 and 2006, more than fifty ships ran aground, sank, or reported damage in South African waters, though not all of these accidents resulted in an oil spill. Fortunately, from 1968 on, SANCCOB has been on hand to rescue seabirds that have been harmed in spills in this part of the world. The following timeline defines the fifteen major oil spills that have affected African penguins residing on islands along the South African coast:

- November 1948: The *Esso Wheeling* sank 30 miles from Dyer Island. It is not known how many gallons were spilled or how many penguins were affected, but the number was in the thousands. There were no penguin rescue centers at the time and it is estimated that one third of the breeding colony on Dassen Island perished in this oil spill.

- August 1952: A spill of unknown origin and unknown size oiled more than 1,200 penguins that came ashore on Robben Island.

- April 29, 1968: The *Esso Essen* ran aground near Cape Point, oiling more than 1,700 penguins. Between 4,000 and 15,000 tons of crude oil was spilled and an estimated 14,000–19,000 seabirds were killed. This is the event that prompted Althea Westphal to found SANCCOB.

- November 1970: The *Kazimah* ran aground on Robben Island, spilling 1,000 tons of fuel that oiled approximately 1,000 penguins.

- February 27, 1971: The *Wafra's* hull fractured and it ran aground 6 miles off Cape Agulhas, creating an oil slick 30 miles long and 5 miles across. There are conflicting reports of how much oil was spilled; some sources say 10,000 tons and some say up to 40,000 tons. Regardless of the amount, 1,216 penguins were covered with the crude oil that entered the ocean that day. Fences hastily erected around the nesting grounds on Dyer Island prevented even more penguins from entering the water and getting oiled.

• March 1972: A spill of unknown origin occurred, which drove 4,000 oiled penguins to seek refuge on Dassen Island.

• August 21, 1972: The *Oswego Guardian* and the *Texanita* collided at high speed in dense fog off Ystervark Point. After suffering a catastrophic explosion that rocked buildings 60 miles away, the *Texanita* sank in just four minutes, losing forty-three crew members. One crew member on the *Oswego Guardian* died. Approximately 500 penguins were oiled from the 10,000 tons of crude oil and fuel oil that spilled from these two vessels.

• July 22, 1974: After six crew members mutinied, the *Oriental Pioneer* continued sailing without the use of radar. The captain set a course that would take him 3 miles off Cape Agulhas, when regulations required ships to be 12 miles away. The ship eventually ran aground at Struisbaai, spilling 200 tons of fuel oil, which covered several thousand penguins. The exact number of penguins oiled is not known.

• August 6, 1983: The supertanker *Castillo de Bellver* exploded and broke in two off Saldanha Bay. Three crew members were killed in the accident, and oily black rain fell on coastal towns in the area for twenty-four hours after the explosion. An estimated 50,000–60,000 tons of the 252,000 tons of oil on board spilled, creating an oil slick 1,200 square miles in size that contaminated 800 penguins.

- August 1985: The *Kapodistrias* ran aground on Thunderbolt Reef at Cape Recife, Port Elizabeth, spilling an unknown amount of oil that oiled more than 1,180 penguins.

- June 20, 1994: The *Apollo Sea* sank near Dassen Island, spilling 2,470 tons of fuel oil that oiled approximately 10,000 penguins. All thirty-six crew members were lost when the ship went down. Half of the 10,000 penguins that were covered with oil perished.

- July–September 1995: An oil slick of unknown origin off Danger Point oiled approximately 1,500 penguins.

- November 14, 1996: The *Cordigliera* sank north of Port St. Johns, with all twenty-nine hands on board. The ensuing spill oiled 1,200 penguins.

- May 1998: An oil supply pipeline burst in Cape Town Harbour. Five hundred tons of oil leaked into Table Bay, with 5 tons ending up in the harbor. In this event, 563 penguins were oiled.

- June 23, 2000: The *Treasure* sank between Robben and Dassen Islands, spilling 1,300 tons of oil. Rescuers removed 19,000 oiled penguins from the two islands. Another 19,500 unoiled penguins were relocated to clean waters 560 miles away to prevent them from becoming oiled as well.

More than 40,000 African penguins were oiled as a result of twelve of these ships sinking. This does not include the birds impacted by the two spills that occurred in 1948 and in 1974, during which several thousand additional penguins (though it will never be known exactly how many) were oiled. Also not included in this figure are the 19,500 penguins that were removed from their islands and relocated during the *Treasure* oil spill to prevent them from becoming oiled. Nor does this reflect the estimated 1,000 oiled penguins arriving at SANCCOB every year that have been contaminated by mystery oil spills. Since its inception more than forty years ago, SANCCOB has rescued approximately 70,000 African penguins plus 13,000 other seabirds, including cormorants, gannets, and gulls. The number of African penguins they have saved is nearly equivalent to the entire population of the species today.

Until the year 2000, the oil spill that had the most devastating impact on African penguins and their rescuers was that of the *Apollo Sea* in 1994. When this aging iron-ore carrier broke apart in a severe winter storm and sank near Dassen Island on June 20, 1994, it spilled 2,470 tons of bunker fuel oil into the ocean. While rough seas are what probably caused the fatal damage to the hull, corrosion and the ship's age (twenty-one years) were believed to be contributing factors in this accident, and both undoubtedly hastened the ship's demise. Whatever the cause, local authorities were not alerted when she ran into trouble. Radio records later showed that, four hours into the ship's voyage, two very brief satellite calls were placed to Beijing; however, nothing ever came of those calls. Perhaps the hull damage was so extreme, and the sinking of the vessel so swift, that the captain never had time to contact rescue officials. No one will ever know. Tragically, all thirty-six men on board were lost as the carrier went down.

Because shipping authorities in South Africa had not been informed when the ship foundered, no one was aware of the loss of the *Apollo Sea* until oil began washing up on shore several days later. Only then was it discovered that thousands of African penguins breeding on Dassen and Robben Islands had come into contact with the oil that had spilled from the sinking ship. But the ship's owners, apparently unwilling to pay for any damage to wildlife and the environment, steadfastly refused to acknowledge that the oil slick was from the *Apollo Sea*. It was only one month later, when the vessel was located via underwater video cameras, that they could no longer deny the fact, and they were forced to take financial responsibility for the ensuing cleanup and rescue effort. The number of penguins oiled in the *Apollo Sea* incident was the largest to date. Ten thousand penguins were rescued–thousands more than SANC-

COB or any other wildlife rescue agency in the world had ever dealt with at once before.

Unfortunately, half of those penguins did not survive. A large number were lost during the rescue and relocation process, due to the unwitting use of poorly ventilated and overcrowded transport boxes and trucks. And because the oil spill was not discovered right away, many penguins were already quite compromised by the time they were rescued, and so died from the effects of being coated in oil for a week–both from ingesting the toxic substance while preening, and from not being able to enter the ocean to feed. Others, still, were lost because they weren't properly stabilized before being washed. SANCCOB eventually called upon other wildlife rescue agencies for help and, together, they did the best they could to save the oiled penguins. But these were extreme circumstances, and no one had ever dealt with a penguin rescue of this magnitude before. The agencies had to utilize whatever previous experience they had managing smaller numbers of animals, and extrapolate from there to try to deal with 10,000 oiled birds. In the end, 5,213 penguins were released after being washed and rehabilitated.

The loss of so many animals was a painful experience for all involved, but many lessons were learned–and immediately after that rescue effort, SANCCOB and local penguin researchers worked together to establish new protocols and guidelines to follow in the event of another large-scale penguin rescue. It was hard to imagine that they would ever have to deal with that many oiled penguins at once again, but they realized that, if they did, disaster preparedness was a critical element in having a successful outcome. To address the issue of bird loss during transport, they designed and manufactured thousands of well-ventilated transport boxes, established a limit on how many penguins could be placed in each box, and planned for the use of open-sided transport trucks in future evacuations. The new boxes were distributed to the penguins' breeding islands so they could be quickly assembled and utilized during the next crisis. That crisis would occur sooner than anyone had dared imagine. Just one year later, more than 1,500 penguins were oiled in a mysterious oil spill; and over a period of four years, a total of 3,265 penguins were oiled and rescued during three separate events. But it was the winter of 2000 that would potentially alter the fate of the African penguin forever.

Exactly six years and three days after the *Apollo Sea* sank, the *Treasure* followed in its ill-fated footsteps on June 23, 2000. Eerily echoing the earlier disaster, the *Treasure* sank just 25 miles from where the *Apollo Sea* had gone down, impacting the same population of penguins on Dassen and Robben Islands, and setting in motion the largest rescue of live animals the world had ever seen. When the *Treasure* first sank, no one could have predicted the size and scope of the ensuing operation. Approximately 75,000 African penguins–41 percent of the global population–were under direct threat from the oil spill, and SANCCOB was on high alert. Ten thousand birds had been oiled in the *Apollo Sea* incident; it quickly became apparent that a far greater number would be affected by the *Treasure* oil spill, and the small center would not be able to manage the rescue and rehabilitation of so many birds alone. The staff needed help, they needed a lot of it, and they needed it fast. But where would that help come from, and would it arrive in time to save the penguins?

5

Disaster Strikes—The *Treasure* Sinks

*Although larger numbers of other species (such as the Magellanic Penguin—*Spheniscus magellanicus*) have been oiled [over time], based on the proportion of the global population which has been affected by oil spills, the African Penguin can be considered the bird species most impacted by oil pollution globally.*

—ANTON WOLFAARDT, CNC AND ADU

In mid-June 2000 the *MV Treasure,* a Panamanian-registered iron-ore carrier, was in the midst of its long journey from Brazil to China when something went terribly wrong. As the seventeen-year-old ship slogged past Saldanha Bay on the west coast of South Africa, and continued southward in rough winter seas toward the tip of the African continent, its steel hull was inexplicably damaged. To this day, no one knows exactly what happened, but an enormous hole suddenly appeared in the *Treasure*'s side, rendering the ship unable to continue its journey. The gaping wound might have been caused by a rogue wave, or perhaps by a cargo container that had fallen off another ship. Floating just below the surface of the water, these containers lie hidden from sight, posing a threat to unsuspecting vessels passing by.

More than 700 miles offshore when the damage occurred, the captain contacted port authorities, asking permission to approach Cape Town to have the *Treasure*'s hull inspected. Though he had a cargo of more than 140,000 tons of ore to deliver, the captain knew he could not continue on to China until he had determined how much damage his ship had sustained. Permission was granted for him to make his way to Table Bay, so the captain steered his 800-foot-long vessel toward this bay, which sits in the shadow of Cape Town's Table Mountain. On Tuesday, June 20, surveyors from the South African Maritime Safety Authority (SAMSA) met the ship outside Cape Town to examine it, but rough conditions at sea made it impossible for divers to do a thorough inspection, and they were unable to ascertain the full extent of the damage. SAMSA directed the *Treasure* to enter Table Bay for a closer look, and on Wednesday, June 21, a more comprehensive inspection was conducted. During this second inspection, a ragged hole, roughly 35 by 90 feet, was discovered in the ship's hull. The question now facing authorities was: What should be done with the disabled carrier? The *Treasure* clearly needed to be repaired, but where and how would those repairs be conducted?

The *Treasure* was too large to be towed into Cape Town Harbour, so SAMSA ordered the captain to leave Table Bay by noon the following day to have the repairs done 50 miles off the coast. The authorities also required that the oil and fuel on the ship be transferred onto another vessel once they were safely offshore. But there were several parties with a vested interest in what happened to the carrier, and valuable time was wasted while these different groups haggled with each other over how to proceed. Key players in the deliberations included the ship's owners (Universal Pearls in Piraeus, Greece); the ship's insurers (Bureau Veritas in Paris); the ship's agents (Good Faith Shipping, also in Piraeus, Greece); the owners of the cargo; and local port authorities.

Although not directly involved in the discussions, closely monitoring the situation were Estelle van der Merwe, centre manager at SANCCOB; local penguin researchers; and conservation officials—all of whom had grave concerns about the welfare of the penguins in the area. They knew too well the devastation that could be wrought by a ship sinking in such proximity to their breeding islands and foraging grounds. Estelle and the others who had been intimately involved with the rescue following the *Apollo Sea* spill knew firsthand how critical it was for proper measures to be taken to prevent the *Treasure* from leaking its oil and fuel into the surrounding waters. The responsible parties in this situation would have to be proactive in preventing another disastrous oil spill from occurring.

It just so happened that, on the same day the *Treasure* limped into Table Bay to have its hull inspected, researchers were out on Robben Island counting penguin nests. Located just five miles outside of Cape Town in the bay, Robben Island is the third largest breeding island for the African penguin. More widely known as the location where Nelson Mandela was held as a political prisoner for the first eighteen years of his twenty-seven-year imprisonment (in the Robben Island Prison), this small island was, at the time, home to approximately 18,000 adult African penguins. During the nest counts on June 21, researchers recorded 1,027 penguin nests with eggs in them and 3,645 nests with chicks in them. Some were slightly older chicks, perhaps four to six weeks old. They had grown in their second, thicker layer of brownish-gray downy feathers, which meant they could regulate their own body temperature and didn't have to be brooded by their parents all of the time. Although these chicks were still in the nest, they were getting too large for the parents to be able to sit on top of them, so they lay next to their parents or jammed their heads underneath their bellies (perhaps believing they could hide from predators that way).

Most of the chicks found that day, though, were tiny fuzzballs, a few days to a few weeks old; these younger chicks just had their original sparse coat of dark brown fluffy down and were still completely dependent upon their parents to shelter them from the elements and predators. Ranging in size from a lemon to a grapefruit, these helpless little chicks spent their days and nights tucked tightly beneath their parents, where it was warm, dry, and safe. Although African penguins lay two eggs at a time, both eggs don't always hatch, and even if they do, both chicks don't always survive. But the 2000 breeding season was shaping up to be the most successful in twenty-five years, and most parents on the island were raising two healthy chicks, bringing the number of baby penguins found that day to roughly 6,000.

As Thursday, June 22, dawned, an agreement about what to do with the *Treasure* had not yet been reached. The noon deadline set by SAMSA came and went, and still the ship languished in Table Bay while the various groups continued their debate about who was responsible and how to proceed. Finally, at 3 p.m., after South African government officials threatened to have the military commandeer the ship, the *Treasure* started to leave the bay under its own power. But the damaged vessel soon began taking on water and was unable to continue any further. At 5 p.m., a helicopter was summoned to airlift the crew from the ship while a tugboat owned by Smit Pentow Marine Salvagers, the *John Ross,* connected a towline to the carrier and began moving it further offshore. At eight o'clock in the evening, when they were 11 miles out to sea, the towline separated. Despite repeated efforts by the crew of the *John Ross* to reconnect the line, they were unable to do so, and they watched helplessly as the *Treasure* began drifting toward shore. Finally, at one thirty in the morning, in a desperate attempt to keep the mammoth ship from ending up on the coast, a helicopter tried to lower two crew members onto the deck to see if they could drop the anchors. But by this time the bow of the ship was completely submerged, and attempting to reach and lower the anchors would have been a suicide mission, so they had to abort the operation.

On Friday, June 23, at three o'clock in the morning, the bow of the iron-ore carrier hit a submerged reef five miles off the coast near Melkbosstrand, a small coastal village known for its white sand surfing beach and the Koeberg nuclear power station (the only nuclear power plant on the entire African continent). Forty-five minutes later, the stern of the ship disappeared below the surface. As the enormous vessel hit bottom, 165 feet down, its fuel tanks burst, spewing 1,344 tons of heavy fuel oil, 56 tons of marine diesel, and 64 tons of lube oil into the waters of Table Bay. At three o'clock in the afternoon of the 23rd, precisely twelve hours after the *Treasure* sank, a three-mile-long oil slick was spotted in the bay near Robben Island.

Although the amount of oil that spilled from the *Treasure* was not tremendous, the environmental impact was monumental. The location of the ship when it foundered could not have been worse–it now lay submerged 26 miles southeast of Dassen Island and 14 miles north of Robben Island, the largest and the third largest breeding islands for the Vulnerable-status African penguin. At the time, these two islands were home to nearly half of the entire world population of these penguins, and the surrounding nutrient-rich waters were their primary feeding grounds. The second largest breeding island for the African penguin, St. Croix Island, lies 560 miles to the east, near Port Elizabeth at Cape Recife. This location would also play a critical role in the coming weeks as the disaster unfolded.

On the same day that the *Treasure* descended to its final resting place in Table Bay, researchers were completing a week of nest counts on Dassen Island and, as they had just observed on Robben Island, it was turning out to be a booming season for penguin chicks. There were 1,075 active nests on Dassen, 420 of which had eggs, and 386 that contained chicks. The chicks in the nests ranged in age from a few weeks to more than two months old. There were also thousands of chicks on the island that were in the final stages of fledging (growing in their first set of waterproof feathers in preparation for independence). These larger chicks had already lost much of their fluffy chick down, and were sporting the sleek silvery-gray feathers of a juvenile African penguin, though they still had patches of fuzzy brown feathers clinging to their heads and shoulders.

Already more independent than the smaller chicks, these young penguins gathered together in groups called crèches, while both parents went to sea each day in search of food. At their larger size, they required much more food to satisfy their increasing hunger, and whenever their parents returned from the ocean with bellies swollen full of fish, they chased them down the beach, begging to be fed. The ravenous chicks whistled and pecked frantically at their parents' beaks until the parents, certain these were their own offspring and not someone else's, finally stopped to regurgitate a meal of partially digested anchovies, squid, and pilchards for them. Thrusting their beaks deep into their parents' open mouths, the chicks greedily gulped down the warm slurry that was brought up. Demanding more and more fish with each passing day, they would soon be kicked out of their nests by the weary parents and forced to fend for themselves. The number of penguin chicks being raised on Dassen Island that week, including these older ones that were close to fledging, was roughly 9,000. In addition to the chicks, about 55,000 adult penguins were present on Dassen Island the day the *Treasure* sank.

The researchers were elated to be finding so many chicks on both of these islands. The numbers of African penguins had been steadily decreasing at a frightening rate for decades, and this positive shift in reproductive success was just what the species needed if it was to survive in the long term. Population models at the time indicated that African penguins could be extinct by the year 2070 unless conditions changed. This breeding season brought a glimmer of hope that things just might be turning around for South Africa and Namibia's beloved penguins. But would this most recent oil spill change all that? Would these birds be able to escape the deadly path of the oil slick, or would the sinking of the *Treasure* set the species back once again? For the moment, the birds on Dassen Island were safe, but the penguins on Robben Island were not as fortunate. The ship had sunk closer to their island, and the oil was already approaching its shores, threatening the survival of both adults and chicks struggling to eke out an existence there. But, at any moment, the winter winds and powerful ocean currents could start moving the drifting oil slick toward Dassen Island, putting the adults and chicks there in jeopardy as well. They were at the mercy of the weather. The potential for this latest oil spill to decimate the species was very real, and immediate action had to be taken if the penguins were to be spared. As head of Cape Town's sole rescue center for seabirds, the future of Africa's only penguin species was now in the hands of Estelle van der Merwe, along with whatever team of wildlife rescuers and penguin experts she could assemble.

6

The Big Cleanup—Oil, Oil Everywhere

It is a curious situation that the sea, from which life first arose, should now be threatened by the activities of one form of that life.

—RACHEL CARSON

After a mostly sleepless night, Estelle was woken by an urgent phone call at dawn on Friday, June 23, alerting her to the early morning sinking of the *Treasure* in Table Bay. Though oil had not yet been spotted near the penguins' islands, she knew it was only a matter of time before it began leaking out of the damaged vessel and threatened a tremendous number of African penguins. This had the potential to become the worst disaster they had dealt with to date–even worse than the *Apollo Sea*–and Estelle was acutely aware that they had to confront it head-on, without a moment's delay. By 7 a.m., she had already spoken with Sarah Scarth from IFAW and Jay Holcomb from IBRRC. She had also contacted Mariette Hopley from the South African Air Force; Mariette had played a fundamental role in the construction of a satellite rescue center for the penguins during the *Apollo Sea* crisis six years earlier. Having worked closely with these three during that rescue, Estelle knew they would be crucial in helping her establish a well-organized, comprehensive, and highly skilled international response. After asking for assistance from these essential players, she contacted SANCCOB staff and volunteers, apprising them of the situation, and preparing them for the worst. Key personnel from the West Cape Nature Conservation Board (CNC), Marine and Coastal Management (MCM)–division of the Department of Environmental Affairs (DEAT), and the Avian Demography Unit (ADU) at Cape Town University were also made aware of the latest turn of events.

While these different groups readied themselves for the probable onslaught of oiled seabirds, the penguins went about their normal business of swimming and hunting in Table Bay. Large numbers of African penguins were gathered in the feeding grounds near Robben Island, repeatedly diving beneath the ocean's surface to capture and swallow swift-moving fish and squid. On each dive, they stayed submerged in the water for three or four minutes before they were forced to come up for air. Floating on the surface between each dive to catch their breath, they unknowingly drifted closer to the oil that was bubbling up from the sunken ship below. The penguins had to hunt for longer than usual in order to eat enough food to sustain both themselves and their young chicks, and were intently focused on their task. But with each passing hour, the threat to the penguins increased.

When the *Treasure* first hit bottom and its fuel tanks cracked, heavy fuel oil trickled out slowly into the surrounding waters; but as the vessel shifted and settled, and was then buffeted by strong underwater swells and currents, the hull sustained further damage, allowing even more fuel to escape. Within a few hours, thick black oil was pouring out in a steady stream and rising to the surface in large globules, where it began to pool together and drift with the shifting tides. By midafternoon, the expanding oil slick was advancing on Robben Island. Soon, it was more than three miles long. Once on the ocean's surface, the oil posed an even greater threat to the penguins, and it wasn't long before dozens of birds began surfacing to breathe right in the midst of the deadly slick. After arriving on Robben Island later that day to assess the situation, Estelle was alarmed to see a few hundred penguins already standing at the shoreline, dripping with oil. She called Mike Meyers, from MCM, and asked him to go out to the island to collect the oiled penguins. He returned to SANCCOB later that evening with about 150 oil-covered birds.

The following day, in an attempt to keep the oil from reaching the coastline of Robben Island, Rick Harding, from MCM's Oil Pollution Unit, set up oil containment booms at the northern end of the island, focusing on the area closest to the penguins' breeding colony. These long, inflatable tubes float on the surface of the water, forming a barrier to hold back drifting slicks. The *Treasure* had sunk north of Robben Island, and the oil, which had drifted south, was now threatening the northern tip where the penguins were nesting. To shield the primary landing area where the penguins entered and exited the ocean on their daily foraging trips, Rick attached the booms to the end of the breakwater at Murray's Bay Harbour. This kept the oil from reaching the penguins' main thoroughfare for several hours, but later that night the booms separated from the breakwater, allowing the oil to reach the rocky shore. Before long, the coastline north of Murray's Bay Harbour was drenched with thick black oil. With the shoreline by their rookery now blanketed with oil, it was virtually guaranteed that every penguin coming from, and going to, sea would be contaminated. Soon, thousands of oil-covered penguins lined the coast, desperately trying to preen the viscous oil from their feathers, but it was no use; it could only be removed by human hands. In addition to trying to save these oiled birds, rescuers would have to try to evacuate all of the clean penguins from the island before they attempted to leave the rookery and enter the water.

There were other serious consequences from the oil drifting in Table Bay. Within a few days of the disaster, several important facilities located on the coastline were impacted. Because the *Treasure* had sunk just five miles from the coastal town of Melkbosstrand, the Koeberg nuclear power plant there was at high risk. When the oil slick first materialized, it was moving toward the power station; fearing the oil would clog the saltwater intake pipes for the cooling system, oil containment booms were deployed to keep the plant operating safely. Fortunately, the winds and currents shifted, preventing the oil from reaching the plant, and a potential nuclear accident was averted. It was the most serious threat the power plant had ever experienced from an oil spill, and the narrowly averted disaster only highlighted the importance of keeping aging and compromised ships from entering the dangerous waters off the coast of Cape Town.

Also affected by the oil spill was the Two Oceans Aquarium, which sits directly on Cape Town's popular Victoria and Alfred waterfront. This facility collects more than 105,000 gallons of water every day from the harbor to fill its tanks, and all of the animals and plants in their exhibits–the sharks, jellies, penguins, frogs, sea urchins, octopi, kelp forests, anemones, and more–were now at risk from the oil floating in Table Bay. Even though oil containment booms had been put into place across the entrance to Cape Town Harbour, oil that had escaped from the *Treasure*'s damaged fuel tanks several days earlier had managed to infiltrate the aquarium's water collection system. Shortly after noon on Wednesday, June 28, a water quality control officer at the aquarium noticed oil in the main receiving tank; by then, it had already clogged filters and contaminated some of the water supply. The water filtration system was shut down while filters were changed and strips of absorbent cloth were put into the tanks to soak up the oil. The aquarium had to switch over to a closed water system, which recirculates the same supply of water through biofilters and aeration units; but this could only be used temporarily, as it was not adequate to properly clean and filter the water for an extended period of time.

Water quality in zoo and aquaria is a very sophisticated science: a delicate balance of pH levels, ammonia levels, chlorine levels, ozone, oxygen saturation, and more must be maintained to ensure a healthy environment for the plants and animals. But until all of the remaining oil had been cleared from the waterfront, the aquarium had to operate on this backup system, which was less than ideal. The longer their normal water filtration systems remained down, the greater the risk to the animals on exhibit. The staff could only wait it out, and hope that the divers attempting to clear the remaining oil from the ship, and water-skimming crews working in the bay, could quickly remove all of the oil. If they couldn't, the situation would quickly become life-threatening for the aquarium's residents.

It was evident that fuel and oil were still leaking from the ship, but no one knew how much remained trapped in the sunken vessel. To deal with the residual oil on the wreck, and to mitigate further damage to the environment, the owners of the *Treasure* and its managers contracted several different companies to manage the cleanup and salvage effort. Divers from Smit Pentow Subsea (the diving division of the local salvors, Smit Pentow Marine), were dispatched to get a closer look at the ship. Smit Pentow also subcontracted divers from Cape Diving and Atlatech Divers & Salvors to help with the salvage operations. They went out into the bay and dove to the bottom to examine the *Treasure,* but strong swells and heavy weather had churned up the water, severely reducing visibility and making it difficult to see the wreck clearly. Divers had to be careful not to get entangled in loose wires and cables that were drifting and dancing about, and they also had to beware of heavy hatches and doors that swung open and shut without warning in the strong underwater surge. They were already working in extremely challenging conditions, but the poor visibility made their task even more hazardous.

While waiting for a work barge and specialized oil removal equipment to arrive from Durban, South Africa, and the Netherlands, divers closed the vents in the ship's fuel tanks, shut down all the fuel valves, and plugged up any holes in the hull where oil was leaking out. They were not using traditional dive tanks, but had an umbilical system that pumped a special helium-rich air mixture down to them. The long hose connecting the salvage divers to the surface included cables for lighting, video, and communication. Another hose pumped hot water into their suits to help combat hypothermia in the frigid 46°F waters. At the depth the ship rested, 165 feet down, they could only work for an hour at a time before they had to surface, going through several decompression stops along the way. The "bends," or decompression sickness, from rising too quickly from a dive, can be fatal, so making these safety stops is absolutely essential. A team of twelve divers was working the wreck, but their progress was being hampered by the poor visibility and rough weather conditions, and there was concern that it was taking too long to remove all of the oil. At this point, the animal rescue team was starting to worry that the oil would not be removed by the time they were ready to start releasing cleaned penguins, which meant the birds would just swim through the oil at sea and get fouled again.

To speed up the process, a decompression chamber was brought to the dive site, and instead of going through all of their decompression stops underwater, divers surfaced after one or two safety stops, then immediately entered the chamber on the dive boat to complete their decompression. This was also done to ease the strain on the divers, who were struggling to maintain the proper depths on their safety stops because of the huge swells and surge. All of this salvage work was being undertaken in perilous underwater conditions; the divers were tossed about like rag dolls by the strong surge, and sharp, ragged edges of torn steel were a constant hazard as they were swept past damaged sections of the hull. One of these powerful underwater swells picked up a diver and smashed him headfirst into the hull. Fortunately, he did not sustain any injuries. The same could not be said for the dive helmet he was wearing. There were a few times when the salvage operation ground to a complete halt because the conditions at sea were just too hazardous and too many lives would be in jeopardy if they continued their dangerous work.

To remove the oil from the sunken ship, a special new "hot tap" system was brought in from Smit Tak in Rotterdam (the parent company of Smit Pentow Marine). This innovative device was designed to drill into fuel tanks underwater without allowing any of the fuel to leak out into the surrounding waters. The *Treasure* had the dubious distinction of being the first shipwreck in South African waters to have this new technology applied to it during salvage operations. A construction barge named the *Winbuild* was towed in from Durban to assist with the operation, and was anchored right above where the *Treasure* lay. Large holding tanks on board the *Winbuild* would store the oil and fuel that was removed from the wreck. With the recovery vessel finally in place and specialized new tools in hand, the divers submerged to continue their work. Using a magnetized hydraulic drill with special cap and valve fittings, divers pierced the steel hull of the wrecked ship, drilling directly into the fuel tank. After the three-inch hole was capped, a control valve was fitted to it, and a long hose running to the surface was attached. Once everything was connected, the fuel was forced up through the hose to the surface by the increased water pressure at that depth. When it reached the surface, pumps onboard the *Winbuild* helped suck the thick fuel into the storage tanks.

Because the *Treasure*'s fuel tanks had cracked as it sank, some of the oil that had escaped was trapped in nooks and crannies on the ship, and it took some time to find all of these hidden pools of oil. Over the course of three weeks, this hot tap procedure was repeated over and over, as dozens of holes were drilled into the ship's hull, and the remaining oil and fuel was removed. The last of the trapped oil was finally extracted from the *Treasure* on Tuesday, July 18. In all, about 200 tons of heavy bunker oil and diesel fuel were recovered. A skimming vessel, the *Albatross,* was chartered from Strategic Fuel Fund to clean up the 1,300 tons of oil and fuel that had previously reached the ocean's surface. Brought to Cape Town from Saldanha on June 27, this vessel could hold just 40 tons of skimmed oil at a time, so the removal of oil from the waters of Table Bay took several weeks to complete.

While it was imperative to remove any remaining oil from the *Treasure,* it was just as critical to remove the oil that had reached the shores of the penguins' breeding islands. If the oil was not cleared from the shorelines by the time the rehabilitated penguins were released, the birds would just get oiled again and have to be rescued a second time. Hoping to prevent this scenario from occurring, cleanup crews were quickly dispatched to their breeding islands. They started cleaning Robben Island on June 26, and by July 1, they had begun work on Dassen Island. The coastline cleanup was coordinated by the Department of Environmental Affairs and Tourism, through the

Marine and Aquatic Pollution Control section of MCM. They, in turn, hired Waste Care's Hazardous Chemical Response Team to oversee the effort. Also assisting in this immense job were the Cape Town City Council, the South Peninsula Municipality, the Teamsters, Portnet, and Cape Diving.

In all, approximately 150 workers were tasked with removing the toxic oil from the coastlines. Many volunteers also turned out to help. Covered head to toe in oil and grime, these workers labored for four weeks, at a cost of $120,000, to clean every bit of the viscous stuff from the rocks and beaches of Cape Town and Robben and Dassen Islands. Large amounts of oil-soaked kelp had also washed up and amassed in piles on the beaches. Teams of volunteers and city workers patrolled the coasts, collecting the contaminated kelp and putting it into large garbage bags to be removed from the area. Each person walked up to 14 miles a day during the cleanup effort, often lugging heavy equipment or bags filled with kelp. Industrial vacuums with huge collection tanks were brought in to help suck up the muck and goo; but the suction would frequently break or the pressure would drop, and workers had to wait while the pumps were primed and pressurized again, slowing down an already sluggish process. The crews often had the exasperating experience of meticulously cleaning a section of the shoreline only to have the tide come in, bringing more oil with it and fouling the area they had just labored for hours to clean. Fighting frustration and defeat, they would return to the same area the next day after the tide had receded and clean it all over again. This process repeated itself over and over during the month-long coastal cleanup on the islands and the mainland.

The industrial vacuums sucked up large quantities of oil, but could not get to all of it, so a local company, Bio-Matrix, was contracted to remove the layer of oil still coating the rocks and beaches. They utilized a new product from Canada, called SpillSorb, to soak up much of the remaining oil. Made from *Sphagnum* peat moss, this natural product has phenomenal absorptive properties. To prepare the moss, it is dried so that only 10 percent of the water remains; after that, it can absorb ten times its own weight in hydrocarbons.

And when used in water, as was done off the coasts of the penguins' breeding islands, it floats on the ocean surface and soaks up all of the oil, fuel, and other chemicals, but not water. The saturated moss product is then removed and buried, where it breaks down in an environmentally friendly manner; the peat moss itself contains humic acid, which is a powerful catalyst in breaking down hydrocarbon molecules. Microbes in the soil produce additional enzymes and the oil is eventually digested, until all that is left is carbon dioxide, water, and fatty acids. On shore, peat-based dust was used to coat the rocks to dry them. After all of the oil-soaked SpillSorb was collected from the water and the rocks, the rocky coastlines had to be painstakingly scrubbed by hand, using wire brushes and detergent. Finally, by July 18, the islands and coastlines of Cape Town were mostly clean.

While these organizations were dealing with the oil at sea, on shore, and on the sunken vessel, rescue workers were faced with the monumental task of evacuating some 40,000 or more oiled and clean penguins from their breeding islands. And something had to be done with the thousands of penguin chicks that would be orphaned once their oiled parents were rescued. During the first few weeks following the oil spill, there were large crews of conservation workers and volunteers on Robben and Dassen Islands working frantically to collect all of the oiled penguins in the crowded rookeries.

Initially, they had to concentrate on collecting the adults, and were forced to leave the small, helpless chicks behind. This was not an easy decision for anyone; however, it was necessary in the midst of this monumental crisis. Though this approach may seem inhumane, these were highly unusual circumstances, and the survival of an entire species was at stake. Researchers knew from many years of studying African penguins that there is a high success rate in rehabilitating oiled adults, and that 86 percent of the adults survive from year to year. More important, oiled adults that have been rehabilitated continue to breed successfully. Data gathered following the *Apollo Sea* oil spill rescue showed that, after being released, rehabilitated adults went on to raise between one and four chicks every year for many years, helping to boost the rapidly dwindling population.

They also knew that not all chicks survive to fledging, and of those that do, only 15 percent survive to breeding age. Forced to make a choice about where to direct their time, energy, and resources, rescuers had to choose the adults. Spending time raising chicks instead of treating the oiled adults, when only a tiny percentage of the chicks would survive to breed, wasn't logical from a conservation standpoint. This decision by the penguin researchers and wildlife rescuers was also affected by their awareness that hand-rearing baby penguins is incredibly time-consuming and requires a certain degree of training and skill. With limited resources and insufficient manpower (and relying almost entirely on untrained volunteers), efforts had to be focused on rehabilitating the adults and returning them to the wild as soon as possible. Because they would resume breeding, these adults were the hope for the future of the species. The teams would return to the islands to collect the abandoned chicks once they had rescued the adult penguins. In conservation management terms, it was a no-brainer. In human terms, and on an emotional level, it was a decision that was extremely difficult to accept.

7

Penguins in Crisis—The World Responds

This single tragedy has done more than any huge marketing campaign could have done to raise the level of awareness amongst South Africans about the importance of the environment, why the environment is so sensitive, and what we need to do.

—MOHAMMED VALLI MOOSA, SOUTH AFRICAN MINISTER OF ENVIRONMENT AND TOURISM

Within hours of the *Treasure* sinking off the coast of Cape Town, heavily oiled penguins began streaming ashore on nearby islands and beaches. Because the oil spill was located close to Robben Island, this was where most of the penguins first made landfall. Some of the birds were from other breeding islands, but had been hunting in the foraging grounds near Robben Island and sought refuge on the closest land they could find after becoming covered with oil. As the icy ocean waters penetrated their feathers and hypothermia began to set in, instinct drove them to head for shore. Before long, the rocky coastline of this small island was crowded with penguins weak from the cold, and exhausted from their strenuous swim through the thick oil. After emerging from the ocean, penguins typically stand near the water's edge, meticulously preening and realigning their dense overlapping feathers to keep them water-resistant. But now, thousands of befuddled penguins slipped and slid on the oil-covered rocks along the coast, as they tried in vain to remove the sticky black substance coating their bodies. It was impossible for them to remove the oil and restore their feathers' usual watertight structure. Unless these penguins were rescued and cleaned, they would all die.

These birds were part of the first wave of oiled penguins that were caught and brought to SANCCOB for rehabilitation. But they would have to wait more than a week before they could start to be washed; it was imperative that the rest of the oiled penguins were quickly removed from the island, as they couldn't go to sea to hunt in their compromised state. The immediate priority was to collect as many penguins as possible and start the process of stabilizing and feeding them; then the crews would begin the time-consuming task of washing them. In the first few days of the rescue, just a handful of people were out on Robben Island catching the penguins and transporting them back to the mainland; Estelle and the SANCCOB staff and volunteers, along with Mike Meyers from MCM and staff from the West Cape Nature Conservation (CNC), did the bulk of the work. Over the course of four backbreaking days, they captured close to 5,000 oiled penguins and brought them back to SANCCOB for rehabilitation.

Corralling and capturing the oil-soaked penguins was no easy feat. While it's easier to catch a bird that cannot evade capture by flying away, catching penguins comes with its own unique set of challenges. Penguins are very wily critters; they're clever and quick and ferocious when cornered. They will not hesitate to lash out and bite and, with beaks capable of deeply slicing into human flesh, they're extremely intimidating to grab. Before long, the hands and arms of people collecting the penguins were crosshatched with dozens of bloody gashes and angry red welts. Not only is catching penguins a very painful task, it truly can be quite dangerous. Like many birds, when threatened, penguins instinctively aim for people's eyes with their razor-sharp beaks. In fact, a few weeks into the rescue effort, Anton Wolfaardt, a reserve manager with the CNC on Dassen Island, narrowly missed having his eye gouged by an agitated penguin. As he was sloshing through the water along the shoreline trying to catch the bird, he slipped on the oil-covered rocks. As he fell, the penguin lunged forward and sank its sharp beak into his upper lip. The wound it inflicted was so severe that Anton had to get several stitches to sew his lip back together. Recounting the incident later, even he had to laugh about the irony of training inexperienced volunteers how to properly catch and handle penguins while sporting a serious penguin bite on his upper lip. There's no doubt that the fresh recruits he trained were more cautious than most after seeing what could happen to someone who actually knew what they were doing when it came to penguin catching.

Estelle, and some of the others with her, had experience catching and handling wild penguins, but most of the volunteers did not. These novice helpers had to put any fear or hesitation aside when reaching into a penguin's nest or into a group of corralled birds. It definitely took courage to plunge one's arms into a swirling mass of penguins guarded by snapping beaks, particularly for those who didn't know what they were doing. One thing is certain; the learning curve was incredibly steep. This was through sheer necessity: if someone didn't quickly learn how to catch penguins correctly, they would be on the receiving end of hundreds of incredibly painful and bloody bites. It was also a numbers game: with thousands of penguins to remove from the islands, catchers got a tremendous amount of practice in a very short period of time. But ask any volunteer their motivation for quickly learning the basics of safe penguin handling, and they'll tell you it was purely a matter of self-preservation.

Capturing and transporting the birds involved logistical challenges as well. In any oiling event, it's critical to find and recover the contaminated animals as quickly as possible. Oiled penguins do something that ends up working to the rescuers' advantage (as well as to their own, ultimately). Once they have been oiled, rather than stay at sea, where they'd be much harder to round up and catch, penguins generally head straight for dry land. This survival behavior makes it easier to locate and capture them. However, when humans show up and try to catch them, if they're near the shore, penguins will instinctively head for the water, as it's the medium in which they can maneuver most easily.

To prevent the penguins from rushing back into the ocean during the *Treasure* rescue, volunteers lined up along the shoreline carrying 10-foot-long sections of portable fencing, placing themselves between the penguins and the water. They then herded the birds away from the ocean's edge and, using additional sections of fencing, maneuvered them until they had a small group corralled. Once they were contained, each penguin was grabbed and dropped into an open transport box, until each box contained three to five penguins. After taping the boxes shut, they were loaded onto small trucks and driven across the island to a ferry that brought them to Cape Town Harbour; from there, they were loaded onto another truck and transported to SANCCOB, where they were admitted. The ferries carrying the oiled penguins were normally used to transport tourists to the island to visit the Robben Island Prison Museum; during the *Treasure* oil spill, these boats ferried thousands of rather unusual–and undoubtedly bewildered–passengers.

By Monday, June 26, three days after the spill, the removal of the oil from the beaches and shoreline of Robben Island had begun, and the wind had finally shifted direction, moving the oil away from the penguins' landing area. At this point, SANCCOB was already bursting at the seams with nearly 5,000 oiled penguins, and it was obvious that thousands more would still need rescuing, but there was no place to put them. To address the urgent need for more space, work began on the creation of an emergency satellite rehabilitation center seven miles away. Although there were still more oiled birds to remove from the island, volunteers collected about 250 clean penguins that day to prevent them from getting oiled. This small group of penguins would be transported the next day to Cape Recife, 560 miles east of Cape Town. After being released into the clean waters there, it was expected they would swim back to Robben Island, a journey that should take two to three weeks. It was hoped that, by then, the oil would be removed from the island and coastal waters, making a safe return possible.

The next day, while the clean birds were on their way to Cape Recife, workers returned to Robben Island to continue catching oiled adults and start collecting the chicks that had been abandoned when their oil-covered parents were rescued. The removal of adult penguins and chicks from Robben Island would continue for another three weeks. Another pivotal event on Tuesday, June 27, was the official opening of the Salt River Penguin Crisis Centre, the temporary rehabilitation facility that would come to house the majority of the oiled penguins rescued during the *Treasure* oil spill. Once this enormous satellite center was up and running, the folks at SANCCOB could breathe just a tiny bit easier. Though they were still tremendously overcrowded and had throngs of oiled and ailing penguins to deal with, at least they no longer had to worry about where they would house the rest of the birds that were still out on the islands. They now had another location where they could bring the impending onslaught of oil-covered penguins.

The pressing need for this massive rehabilitation center would become even more evident in the days to come. While rescuers and researchers had heaved a huge sigh of relief the previous day when the winds shifted and the oil began moving away from Robben Island, they were now all holding their breath because those same winds and currents had begun moving the oil slick in the direction of Dassen Island. Every day since the *Treasure* had gone down, penguin researchers and conservation officials had flown over the area to monitor the location and movement of the drifting oil slick, which had spread across Table Bay like shiny black tentacles on the surface of the water. On their flight that Tuesday, they could see that it now directly threatened the penguin colony living and breeding on Dassen Island. In a desperate attempt to avert the oiling of thousands more penguins–particularly when space, resources, and manpower were so limited–a decision was made to repair a low wall that was already in place around the perimeter of the island, and move as many penguins as they could inside it to prevent them from going into the water and getting covered with oil. But this would just be a stopgap measure. The birds could only be kept confined to the island for so long; there wasn't the manpower to be able to feed them by hand, and without access to the ocean, they would eventually starve. Rescuers could only wait and hope that the oil moved away from Dassen Island soon, so they could release the penguins and let them forage for food in the adjacent waters.

But then, on Wednesday, June 28, everything changed. During a surveillance flight over the islands that day, it became clear that the oil slick was going to hit Dassen. The worst fears of the researchers and rescuers had now been realized. Suddenly, the stakes were dramatically higher. The brief sense of relief they'd felt two days earlier turned to despair. If the oil did not move away from Dassen Island, they would have no choice but to evacuate the penguins. The teams were devastated by this latest turn of events. One researcher, overwhelmed by the news, immediately burst into tears. They already had a massive animal crisis on their hands. Now it had become a disaster of epic proportions. Close to 10,000 of the 18,000 penguins on Robben Island had already been brought to the mainland rescue centers, and now there were some 55,000 penguins on Dassen Island that might need to be removed as well. But where could they possibly house so many penguins? Even the newly built Salt River rescue center wasn't large enough to accommodate this many birds, and there just were not enough volunteers to feed and care for such a huge number of animals.

Stressed-out researchers and rescuers, already at their wits' end, now had to figure out a solution to this additional dilemma. They already had SANCCOB and Salt River to run; nearly 10,000 oiled penguins at the two rescue centers to feed, clean, and rehabilitate; another 9,000 oiled penguins still in need of rescuing; thousands of volunteers to recruit and train; thousands of abandoned chicks to raise; and now, on top of everything else, they had an island full of unoiled penguins that needed relocating. They soon realized that this was going to be a rescue effort unlike any that had come before it. The number of penguins requiring rehabilitation or relocation would eventually exceed the number of live animals handled in any other wildlife disaster in history. Rescuers knew they were in the midst of something monumental and unprecedented. The question was, Would they get the help they needed to be able to pull it off?

8

An Impassioned Plea—Getting the Call to Help

The response has been incredible. We've had people from all over the world coming in to help.

—ROBIN THOMPSON,
YOUTH HOSTEL OWNER IN CAPE TOWN

Although Wednesday, June 28, started off like any other day at the aquarium in Boston, my life was about to change in ways I could never have imagined. Having just finished feeding the penguins and cleaning their exhibit, my colleague Heather and I were standing in front of the penguin office door, buckets of fish in hand, wetsuits dripping on the floor, when the phone inside the office began to ring. Fumbling with the office keys, Heather swung open the door and grabbed the phone. It was Estelle van der Merwe calling from SANCCOB in South Africa. As Heather listened intently, I watched the expression on her face grow increasingly grim. We had been aware of the *Treasure*'s sinking and had kept abreast of the situation online. But now, Estelle explained, there were more oiled penguins at the rehabilitation centers than they'd ever had before, and they did not have enough people to take care of them. And, as of that morning, the situation had just gotten worse because the oil slick had hit Dassen Island, making the rescue of thousands more oil-covered penguins inevitable. They had a crisis of unmanageable proportions on their hands in Cape Town, and they were in dire need of help. Estelle was desperately trying to round up penguin experts from zoos and aquariums around the world to assist with the enormous task of rehabilitating the oiled penguins and training new volunteers.

Though she was relying on the expertise of these skilled professionals to provide guidance regarding penguin health care and help manage the workers during the rescue effort, the bulk of the feeding, cleaning, and care of the wild birds would have to be done by inexperienced volunteers who had never had contact with penguins before. Estelle estimated they would need 1,000 people each day to feed and care for the oiled penguins–but that many penguin specialists didn't even exist (there were just a few hundred worldwide), so the fate of the penguins would have to be entrusted to the good people of Cape Town. These would not be animal specialists or ornithologists; instead, they would have to rely on ordinary folks with little or no animal experience to take care of the penguins. In a city already burdened by widespread crime, homelessness, and AIDS, how significant was a bunch of birds? Would the citizens of Cape Town care enough about these beleaguered animals to want to save them?

Estelle implored us to fly to South Africa as soon as possible to assist with the rescue effort, but the timing could not have been worse. We were in the midst of our own penguin breeding season at the New England Aquarium, and leaving at this time could potentially compromise its success. Raising penguin chicks is incredibly labor-intensive and requires many hands; we had just three penguin staff members sharing the staggering workload. Three penguin chicks had been born within the last month, and several more eggs were due to hatch at any moment. Each of the two chicks being raised by their parents in the exhibit had to be monitored daily and weighed every other day to ensure it was thriving. And, occasionally, inexperienced first-time parents rejected their newborn chicks, which meant that we had to raise them by hand behind the scenes. We already had one Little Blue penguin chick that was being hand-reared. If we both went to Cape Town at this time, it would leave the newest member of our team, Mandy Mirran, to raise the three chicks primarily on her own; and if any more hatched after we left, she would be responsible for their care as well. Though she had helped us raise chicks in the past as a temporary staff member, this was Mandy's first year as permanent staff during the breeding season. In our absence, she would have to make all the chick-rearing decisions alone. This is a tremendous amount of pressure on anyone; raising baby penguins is complex and challenging; even with prior experience, it's beneficial to have other staff members there to discuss concerns or issues. As with most animals, penguins generally mask any health problems until they are quite ill; so by the time a chick exhibits symptoms, things can go downhill quickly.

Mandy would now have to work exhausting fifteen-hour days that entailed preparing a special formula for the chicks every morning; weighing and hand-feeding them every few hours between 6:30 a.m. and 9:30 p.m.; and closely monitoring their health and progress. The first few weeks of life for a penguin chick are always tenuous, so it's stressful no matter how many times you've gone through the process before. In addition to these chick-rearing responsibilities, Mandy would have to manage the care of all sixty-nine penguins in the exhibit and supervise dozens of penguin colony volunteers.

While she was extremely capable, asking just one person to bear the entire workload and assume full responsibility for the penguin colony was an enormous burden. She could call upon the Dive Team for some assistance–all three dive staff members had previous experience raising penguin chicks–but the help they could provide would be limited because of their own duties and demanding daily schedules. The stress and fatigue during chick-rearing season were legendary; even with help from the extra staff that we sometimes hired during these taxing months, we typically worked 65–75-hour weeks! And we worried constantly about the newborn chicks, even while at home. During breeding season (or whenever an animal was ailing or extremely old) my co-workers and I shared the same distressing affliction: we would frequently wake in the middle of the night, our minds spinning with worry about the animals back at the aquarium. We would instantly be wide awake and on edge. It wasn't even a matter of waking up and then consciously starting to think about the penguins–it was as though the racing thoughts themselves woke us from our slumber, tormenting us and preventing us from finding sleep again. It was a running joke that the aquarium completely took over our lives. We all loved the penguins and were intensely dedicated to our work, but it was the type of job that definitely did not stop when you left the building at the end of the day. The wonderful birds we spent most of our waking hours with felt like members of our extended families. Even on our days off, the aquarium and the animals we cared so much about were never far from our thoughts.

Yet, despite the added stress and increased workload our absence would create for Mandy, we knew we could not ignore Estelle's impassioned plea for help. They needed as many experienced hands as they could get in Cape Town, and they needed them right away. The moment she got off the phone with Estelle, Heather began making arrangements to get us to South Africa as soon as possible. She alerted our department supervisor, Dan Laughlin, and an emergency meeting was arranged with the CEO and CFO of the aquarium, our manager and curator from the Animal Husbandry division, and other essential staff, including a representative from the aquarium's Conservation Department. Less than an hour later, we were seated around a long oval conference table, formulating an emergency plan for the aquarium's involvement in this mammoth rescue mission. After describing the unfolding crisis in South Africa and their desperate need for experienced penguin caretakers, Heather outlined what was required to get us there, and what was needed to keep the Penguin Department running smoothly in our absence. Two temporary staff members would have to be hired right away from our volunteer pool to cover our shifts; however, we did not have extra money in the department budget to cover this unforeseen expense. The representative from the Conservation Department immediately offered her department's assistance with this, as well as with some of our other expenses.

There was debate whether one or two of us should go; two of us heading to South Africa meant leaving behind a new staff member to manage everything in the penguin area. Although she would have assistance, leaving just one primary person to raise all of the chicks presented a dilemma; would it be too much work for one individual, and what were the possible repercussions if there were health issues with any of the chicks while we were gone? A discussion about the potential impact on our breeding season at the aquarium versus the risks to the penguins in the wild ensued, and it was determined that the crisis in South Africa took priority, as the future of an entire species was at stake.

Once it was decided that the two of us would indeed go, we returned to our office to get things in order and to prepare for our departure two days later. Heather immediately set about submitting a proposal to the Conservation Department for an emergency grant from the aquarium's Conservation Fund to cover our travel expenses and the salaries for temporary staff. This fund, which is endowed by contributions from individuals and corporations, is set up for the express purpose of saving animals in distress and for carrying out other vital conservation work in the field. While conservation staff had verbally given their support, this formality still had to be taken care of. We then selected two of our most experienced volunteers to cover our shifts, and wrote coverage notes for Mandy.

All of this happened on Mandy's day off, so she was unaware of everything that had transpired until she arrived for work the following morning. Heather and I had both come in early, and as soon as Mandy walked through the door, Heather asked her to join us in the cramped penguin office. Motioning to a chair, she said, "Sit down for a minute. We have something important to tell you."

"What's going on?" Mandy asked, looking anxiously from Heather to me. "Did something happen to one of the chicks?"

"No, the chicks are all fine," she assured her. "It's about the *Treasure* oil spill in South Africa. Estelle called from SANCCOB yesterday, and apparently they've got more than ten thousand oiled penguins to care for, and they're expecting even more to arrive over the next few days. Except for the SANCCOB staff, they don't have any people there with penguin experience, so she's asked if we could come help with the rehabilitation effort."

Mandy now looked even more concerned. "O-kaay," she said hesitantly. "So, are you going?"

"Well," Heather continued, "we met with the higher-ups yesterday afternoon, and they agreed to send two of us, so Dyan and I are both leaving for Cape Town tomorrow morning. We'll be gone for two weeks."

Upon hearing this, Mandy's face instantly went white and her eyes grew wide. All she managed to choke out was: *"Tomorrow?"* and "*Both* of you?" Her shocked reaction was completely understandable. I, too, would have been panicked at the thought of being the only staff member on duty for two weeks—particularly in the midst of chick-rearing season. There truly could not be a more stressful time to be left on your own. But we were confident that she could handle it. She would have to.

Just forty-eight hours after Estelle's call, Heather and I were on an enormous jet, heading across the Atlantic with penguin caretakers from six other facilities in the United States. There had been no time to get the recommended immunizations or anti-malarial medications. Fortunately, though, both of our passports were up to date. Similar scenarios were undoubtedly playing out at zoos and aquariums around the world, as specialists in other countries responded to this massive animal crisis. In the end, 110 penguin caretakers and wildlife rescuers from 59 facilities in 14 countries would fly to South Africa to assist with the rescue. Most of us couldn't be away from our institutions for very long, and it didn't make sense for all of the skilled professionals to be there at the same time—so we went in staggered shifts over the course of the three-month rescue effort. Tom Schneider, curator of birds at the Detroit Zoo and chair for the AZA's Penguin TAG (Taxon Advisory Group, a conservation management group), stayed in communication with the IFAW staff in Cape Town so he could coordinate the alternating schedules of specialists flying in from AZA-accredited facilities.

There was never a doubt in my mind or a moment's hesitation about going to Cape Town to help. But my parents, aware of the high crime rate there, were very concerned for my safety and questioned my decision to travel to a city where one out of every thirty-three people is the victim of a violent crime. My mother had been very overprotective when I was a child; even though I was now a grown woman, I knew this trip would be a source of great worry for her and would undoubtedly cause her many sleepless nights. While I was well aware of the fact that Cape Town had areas where the mugging, murder, and rape rates were quite high, I downplayed the potential danger and assured my parents I would be perfectly safe while there. I didn't fully believe my own assertions, yet I knew I had to go despite any qualms about my own well-being. After all, the penguins needed us. This was an extraordinary opportunity to contribute to an effort that would impact the future of an entire species. How could we not go? Only later would I discover that my parents' concerns had been completely justified.

9

Salt River—
Birth of a Rescue Center

To see the birds in this state is absolutely shocking. The birds are suffering and time is starting to take its toll on them.

—CRAIG VILJOEN, VOLUNTEER AT SALT RIVER

It's hard to imagine just how much space 19,000 penguins can take up. As the oiled penguins were rescued from the islands, they were brought directly to SANCCOB, but before long, the small rehabilitation center was packed far beyond its intended capacity. They were already housing close to 5,000 penguins at a center designed to hold a maximum of 2,000. Estelle and the others knew they might soon have to provide refuge for another 10,000 oiled penguins–and quite possibly more. But there was no place to house them. The situation was critical. The pressure was on to find a much larger building to convert into a temporary rescue center, but locating and preparing an appropriate building was only part of the challenge; Estelle and her newly assembled rescue team would also need a small army of volunteers to staff this second facility and provide specialized care for the impending onslaught of oiled penguins.

A massive warehouse, normally used to repair coal transport

trains, was secured near Cape Town's waterfront to become the temporary rehabilitation center for the penguins. Its original designation was simply "Shed #12," but once it was renovated, this 236,000-square-foot building (over five acres in size) became known as the Salt River Penguin Crisis Centre. As the rescue operation progressed, the plan was to gradually move the penguins that had been washed to a sprawling space adjacent to the warehouse, where they would complete their rehabilitation. This outdoor area, spread over six acres, would eventually have several large swimming pools and more than a dozen enormous holding pens with 1,000–1,500 penguins in each one.

Mariette Hopley, whom most of us working the rescue viewed as the patron saint of penguins, was the driving force behind the creation of the Salt River Penguin Crisis Centre. This superhuman dynamo, who was a semiretired major with the South African Air Force at the time of the *Treasure* oil spill, managed to procure the resources and manpower to get the Salt River facility built and ready for 16,000 penguins in just three days' time. It was not the first time she had worked such a miracle. During the *Apollo Sea* oil spill six years earlier, when she was just twenty-six years old, Mariette supervised the creation of a temporary rehabilitation center to accommodate the 10,000 oiled penguins that were rescued. With that experience under her belt (as well as the smaller *Cordigliera* oil spill rescue in 1996, which she also oversaw), Mariette already knew exactly whom to contact and how to proceed the moment she received the call to help. A no-nonsense woman, with remarkable stamina, strong leadership skills, and outstanding organizational expertise, she worked tirelessly to ensure the rescue center was functional as quickly as possible.

After the *Apollo Sea* oil spill, Mariette, Estelle, and others had the foresight to realize that plans for creating a large temporary rescue center had to be in the works before the next major oil spill or other environmental disaster occurred. The astonishing number of penguins oiled after the *Apollo Sea* sank had taken everyone by surprise, and no one had been prepared to deal with a rescue operation of that size. Painfully aware of the shortcomings of that rescue, they knew what issues had to be addressed and what changes had to take place to ensure a more successful response to the next disaster. Along with local penguin researchers and scientists, they put together an oil spill contingency plan, so that they would be ready if they ever needed to rescue, house, and treat thousands of oiled penguins again.

During the years between these two major spills, Mariette (who was now an IFAW consultant) scouted the region, locating a few potential locations with suitable structures on site. Immediately after Estelle got word that the *Treasure* had gone down, she called Mariette, alerting her to the situation and warning her of the potential number of oiled penguins that might soon be in need of shelter. Within two hours of their initial conversation, Mariette had the Salt River warehouse secured for their use. As it turned out, even with her advanced survey of the facilities and careful strategic planning, acquiring access to one of the three sites she had previously selected was not going to be as simple as making a quick phone call. The first site she contacted that morning was unavailable because it was being used to film a movie–an unforeseen glitch. The second location was undergoing demolition, and the buildings Mariette had set her sights on earlier were in the process of being bulldozed to the ground. In fact, this second location did not even return her urgent phone call until a day or two later.

Fortunately, she was able to reach the people who managed the Salt River railway station, because it was the last of the three places that could possibly house thousands of oiled and displaced penguins. But it was by no means ready for immediate occupancy. The warehouse was being used for storage at the time, and was full of railway cars and freight containers; but after hearing Mariette's compelling plea, the minister of transport and the people at Propnet who ran the shipping division assured her they would have the building cleared out and ready for her to take over in eight hours. (Mariette is exceptionally determined and persuasive–I imagine there are few people who have ever said no to her.) They kept their promise and had the space emptied by that evening. But there still was an extraordinary amount of work that had to be done before animals could be brought in for housing and treatment. Mariette and her construction team were left with an enormous building filled with grime and coal dust, and none of the equipment or supplies they would need for running an animal hospital. Undaunted, she immediately got to work building a wildlife rescue center from the ground up, in preparation for the thousands of oiled penguins that would soon arrive.

All they had to start with were four walls and a roof. Everything else–literally everything–had to be ordered and brought in. Although initially it was equipped with just a few basic utilities, within the space of a few days the empty building was transformed into a fully functioning hospital for the penguins. To achieve this monumental feat, a team of volunteer workers had to construct the following: two washrooms, a drying room with heat lamps, dozens of indoor swimming pools, and a huge storage area for the mountain of supplies. In addition, they needed a triage area for incoming penguins, an intensive care unit (ICU) for very compromised birds, a chick-rearing room, and four massive rooms with hundreds of holding pens to house the penguins; each of these areas was delineated by temporary walls that were made out of enormous heavy-duty black tarps. Electricity, plumbing, and hot water heaters were all installed; refrigeration was brought in to store the penguins' fish; a food prep station was set up for preparing the fish; and a laboratory for running blood work was built. To support the people working the rescue, a volunteer check-in area and a first-aid station were established; portable toilets were installed; and a food station was built where rescue workers could get drinks and snacks.

Mariette and her team installed wiring, lighting, and ventilation systems (for the washrooms only). Then, for washing the penguins, they purchased spray bottles, toothbrushes, washtubs, and garden hoses with special high-pressure nozzles; and for drying them afterward, towels and infrared heat lamps. Cases of detergent and degreaser were donated and delivered. To provide proper medical care for the penguins, they needed medications, syringes, needles, catheters, eye ointment, and vitamins, as well as plenty of fish to eat (and hundreds of food buckets to distribute the fish in). Fire hoses were brought in for cleaning the holding pens, and wheelbarrows and forklifts were obtained for moving sand, shale, and pallets of frozen fish. The volunteers would be required to wear protective gear, so they needed hundreds of pairs of thick rubber gloves, as well as oilskins and Wellington boots. And they too would have to be fed, which–in addition to the food–required thousands of paper plates and cups and plastic utensils. Finally, for the rescue directors and supervisors, tables, chairs, storage lockers, dry-erase boards, clipboards, two-way communication radios, telephones, and computers were supplied. Karen Trendler (who had stepped in as centre manager for SANCCOB during the rescue, so that Estelle could focus on overseeing the entire operation) was responsible for much of the ordering and delivery of these supplies. Staff from Cape Nature Conservation also assumed some of this task.

The warehouse was large enough to comfortably hold four jumbo jets inside, yet it in no way resembled the interior of a pristine aircraft hangar. Instead, the mammoth building was filled with layers of dirt and coal dust from the coal transport trains that had been housed and repaired there. There was no ventilation or air filtration system, save for the dozens of broken windows along the upper perimeter of the walls, so both birds and humans were constantly inhaling small particles of this coal dust. (In fact, our team member, Martin Vince, developed a chronic cough while there, possibly due to this dust. His doctors back in the States couldn't determine an exact diagnosis, but Martin suspected aspergillosis–a respiratory ailment often acquired by birds under a great deal of stress. The disease is transmitted by fungal spores that are found in dirt and in moist environments. Penguins are highly susceptible to it, but humans can become infected as well. Martin's cough hung on stubbornly for six months after he returned from Cape Town (and still crops up occasionally to this day). And, without any ventilation, as the building began to fill with penguins and hardworking people, the temperature crept up until workers were soaked with sweat under their impenetrable oilskins. Railway lines that ran through the building had to be covered with cardboard and thick tarps, as they presented a tripping hazard to people walking through the rooms. Though the space was less than ideal, this makeshift animal hospital would prove to be more than sufficient in providing a safe haven for the oiled penguins while they waited their turn to be cleaned and rehabilitated.

By Tuesday, June 27, just four days after the *Treasure* sank, the building was ready. The first group of 1,400 penguins arrived that day, and during the next six days, another 12,000 oiled penguins were transported to the Salt River warehouse. And to reduce overcrowding at SANCCOB, about 2,000 of the penguins crammed into the small center's holding pens were transferred to Salt River. Eventually, the building would house more than 16,000 oiled penguins. Even before the construction of the temporary shelter was fully completed on that Tuesday, hundreds of boxes of oiled penguins began arriving. Teams of volunteers lined up like a firefighters' water brigade, passing box after box of frightened birds from the transport trucks into the bowels of the enormous building. As the boxes were brought inside, they were lined up in neat rows in Room 1, which was the first room on the left upon entering the warehouse. Before long, the vast space was packed wall to wall with boxes. Teams of experienced penguin caretakers opened each box and quickly evaluated the condition of the birds inside. Once each penguin had been given a cursory exam, it was moved to a holding pool in one of the four rooms–which room it was sent to depended on how badly it was oiled, how thin it was, or how sick it seemed. Penguins that were extremely weak or emaciated were kept in the initial staging area or brought directly to the ICU.

In addition to the thousands of penguins arriving at the Salt River rescue station, there were thousands of Cape Town residents who wanted to help wash, feed, and rehabilitate the oiled penguins. As each person showed up, they were sent to the volunteer check-in area, located just inside the warehouse on the left. A person stationed there took each person's name and phone number, gave them oilskins, thick rubber gloves, and Wellington boots, then directed them to Big Mike, the volunteer coordinator. After assessing their experience and skill set, Big Mike sent them off to whichever room they were most needed in. Once there, the volunteer checked in with the room supervisor to get his or her assignment for their five-hour shift. It was up to each room supervisor to train the volunteers in their area–which, on any given day, could be as many as 250 people. During the early stages of the rescue operation, up to 1,000 volunteers came through the doors of Salt River every day. Every last one of those people was desperately needed to care for the 16,000 penguins in the building.

When imagining what a wildlife rescue operation entails, people often assume it's just about feeding the animals and treating their injuries. In fact, there is so much more that goes into caring for them and ensuring all of their needs are met. The work is dirty, physically demanding, and definitely not glamorous. And an important part of that work is cleaning up after the animals–which essentially means picking up poo. Every day, all of the round vinyl holding pools inside Salt River were broken down and brought outside to be cleaned and disinfected. With up to a hundred penguins in each pool, copious amounts of slippery, malodorous guano covered the floors and vinyl walls. Not only was this unhygienic, but also the acidic guano would damage the penguins' feathers if not removed, so it was vital to clean these pens every day. Several fire hydrants, located a few hundred feet from the entrance of Salt River, came in handy for the continuous task of cleaning these pools. While some volunteers thought this job was "beneath" them or not important enough to warrant their time, many others happily volunteered to wash the holding pens, if only to have the rare opportunity to blast water from a fireman's hose.

The reluctance on the part of some volunteers to do certain tasks was one of the challenges rescue directors occasionally faced. Many of the people walking through the doors of Salt River for the first time wanted the "glamorous" jobs of feeding or washing the penguins. (Even we assumed, when first called to South Africa, that we would likely spend our time doing these tasks–we had not known that we would be asked to help supervise the huge operation. In fact, one penguin caretaker was initially indignant when not given one of these tasks, feeling that his skills and expertise were being wasted. Before long, though, he came to recognize that this attitude was unwarranted.) While the majority of the volunteers were willing to do whatever was needed, we found that there were always a few who needed a little more convincing. With these few, we had to impress upon them the fact that every last chore was critical in ensuring the survival of the penguins; truly, no one job was more important than any other. They all directly contributed to the health and welfare of the birds. If any one link in the chain broke down–be it with fish preparation, cleaning pools and syringes, or feeding the penguins–the birds would not survive. Fortunately, most of the people who volunteered understood this, and gladly did anything that was asked of them.

On the day that Salt River opened, the Red Cross was there to take care of the people working the rescue effort. Their station was set up just inside the main entrance, on the right-hand side. This was where snacks and drinks were distributed to the volunteers, and where their injuries were treated. When they first arrived, Red Cross staff and volunteers probably had no idea they would spend the next few months of their lives stitching up deep wounds from vicious penguin bites, bandaging fingers shredded from force-feeding the birds, and giving tetanus shots to scores of injured volunteers. The more severe wounds ranged from fingers that had been slashed by razor-sharp beaks while feeding penguins to facial injuries inflicted by frightened birds. Most of these cuts and gashes could be treated simply by disinfecting and bandaging them; still, I imagine these Red Cross workers were quite surprised by the amount of suture material they went through during the course of the *Treasure* rescue. Penguins may look cute and cuddly but they are actually quite ornery, and most people are unaware of how powerfully they can bite. Their beaks can split human flesh like a steak knife slicing through butter. And because of their fierce jaw strength, these lacerations can be quite severe and surprisingly painful. After working with penguins for a while, one develops extremely fast reflexes. (To this day, I'm surprised by how effortlessly I catch things as they fall off counters or tables–and I did not have particularly quick reflexes before working with penguins.)

Because we were working in filthy conditions, and the soil harbored the potentially fatal tetanus bacterium, anyone with a wound that drew blood received a tetanus shot. In fact, all of the volunteers and rescue staff who had jobs directly handling penguins were instructed to get a vaccination on site as a precaution, as it was inevitable that we would be bitten by more than one angry or frightened penguin. This requirement led to an interesting encounter between our teammate Alex Waier and a volunteer in his room one day. "I remember one young man, Richard, who complained to me about the tetanus shot he was getting hurting," Alex said. "I called him a big baby, and he said that the first couple shots didn't hurt so bad, but having come to the center ten days in a row now, they were starting to get to him. Ten tetanus shots! I told him to get to the nurses' tent, and then never saw him again. I hope he's okay."

Being wary of needles in an AIDS-ravaged country, I was extremely grateful that I'd been required to get a tetanus shot when I had returned to school eight years earlier to study veterinary technology. Yet the rescue directors still advised me to get a new booster. I'm sure they were using sterile needles and syringes, but I still couldn't shake my doubts, no matter how irrational. Even though my tetanus vaccination was nearing the end of its efficacy, I was more fearful of contracting AIDS than I was of contracting tetanus, so I elected to forgo the booster. (It did cross my mind, more than once, that with so many people bleeding from penguin bites, there was a certain level of risk we all were taking; but I tried not to think about it.)

In addition to being on wound patrol, Red Cross volunteers were charged with feeding up to 1,000 volunteers each day. They made coffee and hot soups, and put together light snacks such as bread with cheese or bologna slapped on it, and PB&J sandwiches. But all of this food had to be procured. There was one woman to whom every person who worked at the rescue centers during the *Treasure* oil spill should be eternally grateful: Vivienne Wolff. A retired fashion buyer and cosmetics executive, Vivienne made it her personal mission to ensure that the rescue staff and volunteers were well fed during the long days and nights at Salt River and SANCCOB. She persuaded many local businesses to donate food and groceries to feed the workers. Supermarkets and food vendors supplied provisions with which the South African Navy made hot soups, and a local bakery donated freshly baked bread every day. Restaurants and catering companies in the area provided hot meals as well. Vivienne coordinated all of these contributions. When all was said and done, $42,000 worth of food had been donated, which provided 100,000 light meals to more than 12,500 hungry volunteers during the fifteen-week rescue. Were it not for Vivienne's tireless efforts, we might have all gone hungry during the *Treasure* rescue.

But even more essential than feeding the people was feeding the penguins. Directly outside Salt River's main entrance, on the right-hand side, was the fish preparation area. This station had tables piled high with plastic bins and buckets, a small space for volunteers to sit while they worked (on top of inverted 5-gallon buckets), and running water with which to thaw the frozen fish. From daybreak until long after the sun went down, this area bustled with activity. A steady stream of volunteers prepared the fish, thawing them and inserting vitamins and medications into the gills of more than 16,000 of them. Every penguin was force-fed three to five fish per day, which meant more than 80,000 fish had to be thawed on a daily basis. (The first fish each penguin was fed each day was one of these medicated fish, and extra had to be prepared because the penguins often destroyed the medicated fish or spat it out.) Other volunteers working this area ran back and forth between the food prep station and the various rooms in the building, delivering buckets laden with fish to the feeders, and bringing back empty buckets–smeared with fish scales and oily residue from the pilchards–to be cleaned and refilled. The 10,000–20,000 pounds of pilchards that arrived each day were stored in large refrigerated units that had been set up toward the left end of the building. As they were needed throughout the day, dozens of pallets of fish were removed from the refrigerators and laid out on the floor of the building or outside to begin thawing.

Ensuring there was enough fish for the penguins presented a challenge in itself. Once the rehabilitation effort was up and running, the penguins at Salt River consumed approximately 5 *tons* of pilchards every day, and during parts of the rescue, they were eating up to 10 tons daily. (That's 50,000 to 100,000 individual fish every single day.) In addition to the pilchards eaten by the birds at Salt River, the penguins at SANCCOB were downing between 1 and 2 tons of fish every day. During the first two months of the rescue effort, the penguins consumed 400 tons of pilchards–that's approximately 4 million fish! But this soon became a problem. One month into the rescue effort, just after Heather and I left South Africa, the fish vendors ran out of their stocks of frozen pilchards and the local fishermen had reached their daily quotas of how many fish they could legally catch. They were already delivering all of the fresh pilchards they caught to Salt River and SANCCOB. Because of the dramatic decline in the pilchard population in the Western Cape area, local fishing restrictions had previously been put into place in an attempt to increase the numbers of these fish. Originally established in an effort to help save the African penguin, these restrictions could now harm the very animals they were intended to help. The situation at the two rescue centers had become desperate. Without fish, the penguins would quickly starve to death, and all of our efforts would have been wasted.

Faced with this new crisis, the rescue directors called for a meeting with the minister of environment and tourism Mohammed Valli Moosa and informed him of their urgent need for more pilchards. Once made aware of the huge numbers of penguins that could die as a result of this food shortage, Valli Moosa agreed to temporarily lift the restrictions on the pilchard catch. Deeply concerned about the survival of South Africa's beloved penguins, he granted special provisional permits that allowed local fishermen to catch as many pilchards as they could, with the explicit understanding that they could not sell the surplus catch. Instead, any additional fish they caught over the previously established quota had to be donated directly to the rescue effort. They readily agreed, happy to do something to help. In fact, a friendly competition soon broke out between the local fishermen to see who could catch the most fish to feed the oiled penguins languishing in the huge shed at Salt River.

Receiving fresh, unfrozen fish, however, had both its benefits and its drawbacks. On the positive side, food preparation was much faster and easier; the quality and texture of the fish were perfect; and none of the vitamins or nutrients had been lost (as happens during the freezing process). However, because the fish were arriving fresh from the boats, they weren't packed neatly in pallets, like the frozen fish. Instead, they were delivered in huge buckets, packed in water. It became a problem when the number of pilchards arriving each day started surpassing the amount the penguins were eating daily. Because the fish were packed in water, the quality suffered if they sat too long; so the strategy had to change once again. At this stage, the fishermen were told precisely how many tons of pilchards were needed each day, and were asked to catch and deliver just that amount. Due to their enthusiastic participation in this venture, the penguins had a steady supply of fish throughout their stay at the rescue centers. A reconciliation took place with the ship's insurers the following year, during which all remaining costs from the rescue effort were covered. From this final settlement, the fishermen were paid for the pilchards they had donated to the cause. In the end, it was a win-win situation for both the penguins and the fishermen.

10

The Grueling Work Continues—The Emotional and Physical Toll

Hope begins in the dark, the stubborn hope that if you just show up and try to do the right thing, the dawn will come. You wait and watch and work: You don't give up.

—ANNE LAMOTT, *BIRD BY BIRD*

My second day in South Africa began with the sound of Heather banging loudly on my bedroom door. Although I had just been in the deepest sleep of my life, I'd had no awareness of it at all. Sleep had come on so fast and hard that, when I first heard her insistent knocking, I thought it was still one thirty in the morning, and was annoyed that she was disturbing me while I was trying to go to sleep.

"Get up!" she called out.

"What for?" I asked.

"Because it's six o'clock. It's time to get going."

"What are you talking about? I just got into bed."

"It's time to get up," she repeated.

"It can't be. I haven't gone to sleep yet." By now, I was rather irritated. Glancing toward the bedroom window and seeing that it was still dark outside, I added gruffly, "Heather, it's still nighttime. Go back to bed."

"No it's not," she replied more forcefully. "It's six a.m. and you have to get up *now*."

I was terribly confused by this exchange, but upon looking at my watch I realized that four and a half hours had just passed in what quite literally felt like five seconds. It was the same sensation I'd had during several surgical procedures when I was younger. Anyone who has had general anesthesia will recognize the total lack of consciousness that I experienced that night. It's an entirely different sensation from being asleep. For those who haven't "been under," there is a surreal sense of emptiness and oblivion accompanying anesthesia that is extremely disquieting. Time instantly stops and consciousness is completely suspended. It feels as if you have entered a vacuum utterly devoid of space and time; there is just nothingness. And then, just as suddenly, there is awareness again. I've often imagined it's what death might feel like. The overall feeling is extremely disorienting and unsettling.

This was the exact sensation I had on our second morning in South Africa, and on every morning during our eighteen days there. My exhaustion was so profound that I did not sleep at all in the normal sense during the rescue operation. Instead, I instantly plunged headlong into unconsciousness every night the moment my body hit the mattress. On occasion, lacking the energy even to remove my clothing, I would lay down "for just a second" before attempting to undress, only to wake to the shrill ringing of my alarm a few hours later in the same position, still wearing my soiled and stinking clothes. As a result, I never woke feeling rested, and my time in Cape Town felt like one interminably long, grueling day. This bone-numbing exhaustion was experienced by everyone involved in the rescue effort, particularly those of us in supervisory or managerial roles, as we worked constantly day and night for weeks on end without stopping. But we had to ignore and push through the relentless fatigue that plagued us, not only on a physical level but mentally, emotionally, and spiritually as well. For those of us who had arrived from zoos or aquariums, the backbreaking work lasted for just a few weeks; but for the rescue directors from IFAW, IBRRC, and SANCCOB, there was no end in sight.

There was only one exception to this wearying pattern. It followed a day when I had one or two Red Bull energy drinks. The beverage company had donated cases of the highly caffeinated, sugar-packed drink to be given out to the rescue workers–which, in theory, was wonderful. The extra energy certainly was helpful in getting us through our demanding day. However, it made falling asleep that night a rare challenge. That was the only night during our stay in South Africa that I did not immediately fall into unconsciousness but, instead, lay wide-eyed and alert for about forty-five minutes after crawling into bed, incredulous that a product existed that could prevent me from falling asleep in my exhausted condition. It was the first–and last–time I ever had a Red Bull.

That second morning, after accepting that it was indeed six a.m. and I had to get out of bed, I pushed up into a sitting position and took stock of my body. It wasn't pretty. My hands and arms were covered with dozens of cuts and bruises from razor-sharp penguin beaks and flailing toenails, and my entire body ached from having spent fifteen solid hours hunched over in the holding pens force-feeding hundreds of uncooperative penguins. Wearily, I climbed out of bed and pulled on some clean bluejeans and my blue New England Aquarium sweatshirt. With the stench from the warehouse permeating the clothes I had worn the day before, I was glad I'd packed an extra pair of jeans and several shirts. I opened the front door to our hotel room and put on my hiking boots. Due to the smell, I'd had to leave them, along with my soiled clothes from the previous day, outside. I was relieved that we had an outdoor entrance to our suite; at least the smell wouldn't offend any neighboring hotel guests. After putting on my boots, I took my filthy clothes to the front desk and, apologizing profusely, asked to have them laundered. The concierge, as politely and gamely as she could, took them into a back office and closed the door tightly behind herself, insisting it was no problem. Embarrassed about foisting my reeking clothes upon the hotel staff, I mumbled a thank-you and slunk away to join the rest of our team in the adjacent restaurant for breakfast.

Surveying my colleagues, I was not surprised to find everyone looking rather bleary-eyed and pensive. I think we all were feeling anxious and a bit uncertain. The previous day had been overwhelming and disturbing. Nothing can truly prepare you for the experience of walking into a warehouse jam-packed with 16,000 traumatized, oiled-covered penguins, knowing that every last one of them has to be force-fed, scrubbed clean, and nursed back to health–and you are expected to play a significant role in making it all happen. I imagine everyone was still adjusting to the idea that, as of today, we were charged with overseeing many aspects of this rescue operation: from chick rearing and food preparation to training volunteers and managing rooms with thousands of birds in them. The responsibility was enormous. After just one day at the rescue center, they were counting on us to help run things, and with very little instruction. Although we exchanged a few stories over breakfast about our first day at Salt River, everybody seemed preoccupied, and there was much less chatter than I'd expected. Everyone tried to eat a good-sized meal that morning, knowing there would be little or no opportunity to eat for the rest of the day–at least until we returned to the hotel after eleven that night. Before we knew it, it was time to climb into the minibus for our twenty-five-minute ride to the rescue center and the waiting penguins.

Upon our arrival at Salt River. I started gagging and dry heaving again as soon as we entered the warehouse. I had mistakenly thought that I might have adjusted to the odor by the second day but, as I was to learn, there was just no getting used to it, and I spent the first few hours of every day breathing through my mouth until I acclimated to the revolting smell. Like an invasive parasite, it infiltrated every pore of our skin, penetrated every strand of our hair, and permeated every fiber of our clothing, backpacks, and shoes. Each night when I showered, as soon as the hot water hit my long hair, the putrid odor locked into each strand was released, filling the shower stall with a heavy mist that smelled like the inside of the warehouse. Even alone in my hotel bathroom, I couldn't escape the stench. The nauseating odor in the rescue center actually drove several volunteers away, never to be heard from again. Overall, the volunteers were exceptional. They were hardworking, compassionate, and dedicated, but there was only so much some of them could endure–and I can't say I blame the few who fled in horror. Between the smell and the heartbreaking condition of the penguins, the working environment was pretty dreadful, and it took a strong stomach and a great deal of resolve to handle it.

Having spent our first day at the rehabilitation center feeding penguins and getting oriented, upon our arrival that second morning we had to quickly work out how we were going to accomplish the tasks we had each been assigned the night before. It was up to each of us to organize systems and establish routines in our areas to enhance the operation and keep it functioning smoothly. But it was essential that our team work closely with the rescue directors to understand the overall objectives and to determine the most efficient procedures for animal care. Our second day–and each day that followed–began and concluded with a briefing meeting with all of the rehabilitation managers and supervisors in an area appropriately called the "War Room." Jay, Linda, and Sarah typically ran these meetings, and when they weren't completely inundated with logistics, press conferences, and other duties, Estelle and Mariette participated in them as well. We addressed any issues or needs we had, and discussed the rehabilitation strategy, which could change daily based on several factors: how quickly or slowly the washing of the birds was going, how many penguins were still getting force-fed versus how many were free-feeding, how many penguins were getting swum daily, how many volunteers they thought might show up that day, and so on. Everyone's input was taken into account; however, the big-picture decisions were made by the team leaders. Our job was to figure out how to achieve the daily objectives, and to make sure they were met before we left each night.

Following our meeting each morning, we went directly to our assigned rooms and checked on the birds in every pen. Oiled penguins were still being rescued from their islands and brought to Salt River throughout our first week there, and new birds were sometimes added to the pools in our rooms after we had left for the night. Some penguins were more profoundly oiled than others. If we found penguins on our morning rounds encased head to toe in a thick black coat of oil, we separated them out into another pen with birds that were slated to be washed as soon as possible. If any of them looked particularly weak or emaciated, we brought them to the veterinarian in the ICU. There were eighty-three pools in our room, fifty-one of which were filled with birds; each occupied pool held between 70 and 120 penguins, bringing the total in our room to 4,500 or so. The thirty-two empty pools were to transfer penguins into once they had been fed, so that the pool they had been in overnight could be cleaned.

Because there were fewer empty pools than full ones, every day a carefully choreographed routine of feeding the penguins, swimming them, and cleaning their pens took place. After each penguin was fed, it was deposited into a nearby clean pool. All of the penguins within a pool were transferred together so that, for the most part, the same groups stayed together until they were washed. A small cardboard tag was hooked onto the wire frame of each pool; when a pool full of penguins had been fed, it was recorded on the tag, along with the date. The tag was then transferred to the new pool that the newly fed penguins were in. Usually. On occasion, no one would remember to move a tag, and we then had to try to figure out where the tag belonged. This was important because we didn't want the same birds to be force-fed twice in one day, and we also wanted to be sure that every penguin was fed each day.

Trying to keep track of which pools of penguins had been swum, which had been fed, and which pools to transfer penguins into after they were fed was a huge logistical headache. After all, a few days earlier, we had just 70 penguins under our care; now, we had more than 4,000. We were using notepaper and clipboards to record the status of each pen, but we soon realized we needed to come up with a more organized system to manage the complicated schedule for the care of so many penguins. So we requested a dry-erase board to chart the eighty-three pools in our room, noting each day which of the fifty-one pools had penguins in them and which of the thirty-two pools were empty. As birds were fed and swum and transferred to new pools, and as dirty pools were cleaned, we tracked those changes on the dry-erase board. It was a rudimentary but effective way of monitoring the status of all the penguins in our room. I felt like a football coach with a playboard, working out the plays for the next big game. But it helped us to make some sense of that enormous room.

By the time we arrived in Cape Town, nine days had elapsed since the sinking of the *Treasure,* and Salt River had been up and running for just four days. It was very early in the rescue mission, and procedures at both Salt River and SANCCOB were still being established and put into place. One of the biggest challenges faced by rescue organizers was finding enough people to help care for the penguins. In a wildlife disaster, one hopes to have plenty of experienced handlers available to care for the animals, but in this case that was not possible. With 16,000 penguins under one roof, and 3,000 under another, approximately 1,000 animal caretakers would be needed on a daily basis. We had no choice but to rely on untrained volunteers; essentially, the lives of 19,000 oiled penguins would be in their hands, and it was up to us to train them. These were people with big hearts and willing bodies, yet none of them had a lick of experience handling penguins. Anyone who has worked with these hardy seabirds can attest to the fact that they are very strong and very cantankerous–and they won't hesitate to bite the heck out of you any chance they get.

Hundreds of brand-new, inexperienced volunteers arrived at the rescue centers every day and each one had to be systematically trained from square one how to properly clean, restrain, force-feed, and swim these irascible birds. Our primary job in Room 2 was to maintain the health of the penguins while they waited their turn to be washed. To achieve that goal efficiently and safely, we had to put our volunteers through an accelerated, comprehensive training program. Some volunteers were responsible for swimming the penguins, while others were tasked with cleaning the holding pens or being "fish runners" (making sure the feeders always had a full bucket of fish next to them). Some volunteers were perfectly happy with these assignments, but others were anxious to learn how to feed the penguins. Before they could do that, though, they had to train to become handlers. As a handler, each had to know how to properly capture and restrain the birds so that neither the handler nor the penguin (nor the people near them) would be injured.

To ensure everyone's safety, we insisted that our volunteers follow the "18-inch rule"–a mandatory regulation for our penguin colony volunteers back at the New England Aquarium. This meant that their face always had to be at least 18 inches from the penguin's face; any of our exhibit volunteers who failed to obey this directive were fired. It was for their protection, after all; those powerful beaks were capable of inflicting a great deal of damage. Enforcing this rule in our massive room at Salt River proved much harder than enforcing it in our exhibit back in Boston–with so many people to monitor in a room longer than a football field, it was nearly impossible–but we still tried our best. Once volunteers had mastered the proper handling technique, they were trained how to carry out the complex process of force-feeding the penguins. Due to the sheer number of birds moving through the system, the learning curve for the volunteers was incredibly steep. Indeed, by the end of the rescue effort, many of them probably had more direct hands-on contact with penguins than caretakers in some zoos and aquariums do!

Though we were confronted with the distressing condition of the penguins every moment of every day, we had to consciously block out the pain of being surrounded by thousands of helpless and suffering animals. Every one of us on the rescue team has an abiding love and concern for animals, but to get through each brutal day, my colleagues and I had to partially shut down emotionally. Otherwise, it would have been just too much to bear. Several volunteers were completely overwhelmed by seeing thousands of animals in such appalling circumstances. They would enter the rescue center and just stand there, looking around in stunned silence at the hundreds of pools overflowing with oiled penguins. Shaking their heads, tears streaming down their faces, they all softly murmured the same thing; "Ach, shame . . . shame." (In their lovely South African lilt, this sounded like "Ach, shemm . . . shemm.") For some, the pain was just too great and, though they wanted to help, they found they could not return to the rescue center to volunteer again.

As for us, with the amount of work that had to be done, we did not have time to stop and truly take the whole thing in. We just ran on autopilot. We also intuitively realized that if we started thinking about how terrible it was, we might be overcome with anger and sadness: anger at the people responsible for this tragedy, and sadness for the animals and what they had to go through. It wasn't until we left South Africa–and the adrenaline had stopped coursing through our veins–that we were able to process the experience and the strong emotions that we had suppressed while there. But there were some moments in Cape Town when reality slapped us in the face, and we were reminded again of the devastating impact of this oil spill.

Making the morning rounds of our room a few days into our stay, I noticed that something didn't look right in one of the pens. Bending over to get a closer look, I realized there was a dead penguin lying on the bottom of the pool. When I picked it up, its body remained frozen in the same prone position; rigor mortis had already set in. Its body was rigid and cold; its eyes open and unblinking. Hundreds of the penguins had arrived at the rescue center in extremely poor condition; some were thickly coated with toxic oil, while others had not eaten for several days, and were already emaciated, weak, and starving. With approximately 4,500 oiled penguins in our room, we could not expect every one of them to survive. Still, it was disheartening to find that this poor penguin had not lasted long enough to have the oil washed off its body, and I felt that I had failed in my duties. Intellectually, I knew it was unrealistic to think we could save every last bird in our room, but emotionally, I still felt a pang of guilt as I lifted the cold, stiff body of this one penguin out of its pen.

During our rounds on another morning later that week, I came upon something completely unexpected: on the bottom of one of the holding pens, there sat a glistening white egg. I felt a momentary flicker of excitement before reality quickly set in. Obviously, this egg would never produce a penguin chick. There was no penguin sitting on it. The egg had been abandoned by its mother after she laid it, and left to perish. Clearly, this was no place to raise a baby penguin; there was no nesting material, no burrow, no ocean, and no protection. And it was highly unlikely that her mate was even in the same pen to share incubation duties, as most pairs had been separated from each other during the evacuation of the islands. (Indeed, out of the thousands of penguins in that building, I only saw a handful that appeared to be mated pairs; these were the few that I witnessed preening each other.) Instead of wasting energy incubating the egg and attempting to raise a chick in unsuitable conditions, this penguin mother had done the only sensible thing, and abandoned her egg. That discarded egg was a jarring reminder of the long-term consequences of this oil spill, and the bearing it would have on the future of the species.

What had been the best breeding season ever recorded for this vulnerable species had been destroyed–literally overnight–by the sinking of an aging iron-ore carrier. When the oiled penguins were rescued from their breeding islands, thousands of eggs and newborn penguins had to be left behind; the deserted eggs would never hatch and thousands of abandoned chicks would not live to see another day. They would slowly starve to death, or they would be picked off by predatory birds, snakes, and cats out on the islands. The lucky ones would be collected and humanely euthanized before they met either of these gruesome ends. Though 3,350 chicks were eventually rescued to be hand-reared, the rest would not make it through this breeding season. Nearly an entire generation of penguins would be lost in a heartbeat due to human negligence.

There were bound to be emotional consequences of dealing with such an unrelenting rescue effort, but one particular incident evokes feelings of guilt and remorse even to this day. Early one morning, just a few days into our stay, a visibly distressed volunteer approached me with an extremely weak and emaciated penguin under her arm. The exhausted bird could not even hold up its head, and its wings and legs dropped limply as she carried it across the room. She asked me to try and save it, so I placed the bird in a transport box and assured her I would bring it to the ICU as soon as I had a free moment. The penguin lay flat and motionless on the bottom of the box, and I immediately knew it would have to be euthanized, but didn't tell the volunteer to spare her the painful truth. It was too far gone to be able to save in this situation.

The unfortunate reality was that, in a disaster of this size and magnitude, every life could not be saved. A herd health approach was necessary, particularly early on in the rescue effort, when thousands of heavily oiled penguins needing urgent care were arriving at the center every day. We were in crisis mode, and severely compromised birds could not be given the intense individual attention required to save them. Dedicating large amounts of energy to one gravely ill animal, when that time and effort could be more fruitfully spent caring for several individuals with a higher probability of making it, was a luxury the species could not afford. This wasn't just about saving the oiled penguins in the building–it was about saving a species that was already struggling to survive.

These were never easy decisions for anyone to make, and it seems harsh, but the few had to be sacrificed to ensure the survival of the rest. As rehabilitation supervisors, we were well aware of this fact; however, we did not share this knowledge with the volunteers, as we knew it would be much harder for them to accept this painful reality. They were working so hard and with so much heart, and it would have been demoralizing for them to know that some of the birds they brought to us would have to be euthanized simply because we didn't have enough time or manpower to treat them. This is not to say that none of the ailing penguins were given medical care–many of them were; and as the rescue operation progressed and things became less hectic, even more could be treated.

As for the penguin the volunteer brought me that morning, I had every intention of taking it to the veterinarian as soon as possible, but was never able to find a moment to leave my post to do so. Every hour or so, the same volunteer came over to ask if I had taken the penguin to the ICU yet, and each time, I had to tell her I hadn't. She was getting more and more distraught, and I was feeling horribly guilty because I realized this penguin was slowly dying, and my delay in getting it to the vet was only prolonging its suffering. Eventually, despite the relentless chaos that held me captive in that room, I forced myself to leave long enough to carry the lifeless bird on its final journey to the ICU. Of the thousands of birds entrusted to our care, this one still tugs at my conscience more than the rest. Even ten years later, I am still haunted by the memory of it.

While we were trying to bring order and routine to Room 2, the other members of our team had their own challenges to deal with. Jill Cox was busy in Room 4 with Mike Short doing essentially the same thing as Heather and I: training hundreds of volunteers, and making sure the 3,000 penguins in their room were fed and swum every day. Gayle Sirpenski had been given the responsibility of preparing all of the fish for the penguins. Martin Vince was helping her, and a large part of their assignment was to train the volunteers who were working at this task. One of the problems we had encountered during our first day at Salt River was the poor condition of the fish that were being fed to the penguins; much of it was not properly thawed, which led to feeding and health issues. Some of the pilchards were overthawed, making them too soft and mushy to easily feed to the penguins. Because we had to literally force fish down the penguins' throats, this soft fish caused a lot of problems. It was so mushy that it would start to fall apart in the midst of feeding, and many of the penguins were breathing in the blood and guts that squished out while the fish was still lodged deep in their mouths. This led to cases of aspiration pneumonia at both Salt River and SANCCOB, so it was critical to revise the preparation procedures to make sure the pilchards were not thawed for too long. A second problem caused by the disintegrating fish was that the oil from their innards got smeared all over the feathers on the penguins' faces and necks, creating an even bigger mess for the bird washers to deal with.

On the flip side, much of the fish we had received in the rooms that first day was underthawed: still icy and almost solid to the touch. I took to calling these stiff, frosty pilchards "fishsicles." After a penguin had a few of these pilchards shoved down their throats, they would stand in the pool shivering violently for an hour or more while the ice-cold fish thawed out inside of their bellies. This shivering was burning calories and utilizing precious energy they could not afford to waste, so an adjustment in the thawing procedure was made to address this issue as well. Within a day or two of Gayle and Martin retraining the volunteers in the fish prep area, we were getting perfectly thawed fish in our rooms.

With 16,000 penguins at Salt River and 3,000 penguins at SANCCOB to feed every day, an enormous amount of fish was needed. Gayle and Martin each picked up and lugged around thousands of pounds of frozen fish on a daily basis. Not only did the pilchards have to be properly thawed, but they had to have vitamins and medications inserted into them. The vitamins that are normally in the fish leach out during the freezing process, so vitamin supplements had to be given to make up for those lost. And because the penguins were suffering from anemia from ingesting toxic oil (the oil breaks down their red blood cells), they also got iron tablets. The easiest way to get pills into a penguin (or almost any animal) is to hide them inside of their food. To put pills into a dead fish, the gill cover on the fish is lifted and the pills are inserted. We use this same technique at the New England Aquarium–most zoos and aquariums do–although, as cats or dogs often will, our penguins seemed to catch on to this tactic, and many would turn their nose up at a fish that had vitamins or medications in it, probably because they could smell the pills inside.

Another one of our team members, Alex Waier, had his hands full training a crew of people to give every penguin in the building an activated charcoal solution to counteract the effects of oil ingestion, and an electrolyte solution (called Darrows) to rehydrate them. Penguins get their hydration primarily from the food they eat, and some from drinking seawater while they're swimming, but by the time they arrived at the rescue centers, most had gone several days without eating. While penguins can survive for many days without solid food, like any animal, they will quickly die from dehydration, so getting some fluids into these birds right away was essential. The birds received these fluids by inserting a tube into their stomachs, through which 50–60 mls of the activated charcoal and 120 mls of Darrows were administered. This is a task that requires skill and precision, so Alex handpicked a team of reliable volunteers who were already trained in handling and force-feeding, and taught them the process. He and his team then went from room to room, tubing every single penguin in the building. Most days, they managed to tube between 3,000 and 4,000 birds; on their best day, they did 5,000. Alex even had the unique and rather surprising experience of having a penguin lay an egg on his lap while he was giving it fluids!

This supplemental hydration was critically important in the early days of the rescue, when the numbers of penguins far surpassed the number of volunteers trained to feed them, and the birds were only getting fed every other day. This was something that bothered me terribly. We had not been able to get every penguin in our room fed the first day or two that we were there; there were just too many birds and too few trained volunteers. It wasn't until a few days later that we learned this was the case in every room. In fact, Jay, Linda, and Sarah knew that it would be impossible to feed every bird daily in the beginning. I hadn't realized this and felt very guilty about failing to do so, and not just because I believed we had let the rehabilitation directors down. What troubled me far more was the thought of some of the penguins going hungry for a day.

While the rest of us were inundated with oiled penguins and truckloads of frozen pilchards, the last two members of our team, Lauren DuBois and Steve Sarro, were up to their eyeballs in penguin chicks. Although many of the chicks had died out on the islands before they could be rescued, and others had to be euthanized, about 3,350 had eventually been saved. These chicks were being hand-raised at several different locations in and around Cape Town, including a large group that had been brought to Salt River. In a separate room set up at the far right end of the warehouse, 723 penguin chicks were getting their second lease on life. Because feeding penguin chicks is a very delicate undertaking, Lauren and Steve carried out all of the feedings themselves, although they occasionally had a volunteer or two help with the free-feeding once the chicks were a bit older. On a good day, they were lucky to have three or four volunteers helping them by cleaning the chicks' pens, but for the most part they were on their own in the chick room. Because this area was not packed full of oil-covered penguins and sweat-covered volunteers, the smell was not as overpowering as in the rest of the building, and mixed in with the familiar stench of the warehouse was the sweet smell of penguin chicks.

Penguin chicks have a unique and unmistakable smell to them—sort of a musky, dusty, fishy aroma (with a hint of guano rounding it out). Missing the constant physical contact they once had with their parents, these little fuzzballs will often nuzzle up against their human caretakers for comfort and warmth; I guess in a storm, any port will do. During each breeding season at the New England Aquarium, we grew accustomed to the penguin chicks melting into us after we fed them; and when they snuggled up under our chins, their unique scent wafted up our nostrils, staying with us for the rest of the day. For the duration of each breeding season, the distinctive smell of baby penguins clung to our clothing and hair, trailing behind us wherever we went. It's such an unusual scent that there's really nothing else to compare it to; but mention "penguin chick smell" to anyone who has raised these adorable little bundles of fluff, and they will instantly nod their head in recognition, a smile spreading across their face. One of my former colleagues from the New England Aquarium actually calls the smell "intoxicating." I can't say I would choose that particular word to describe it—but to each his own.

Because penguin chicks are so darn cute, the chick-rearing room was a place that rescue supervisors occasionally escaped to when they needed a short break from the chaos in the rest of the building, and a reprieve from being surrounded by thousands of oiled penguins and hundreds of people. Years later, Big Mike (who, as volunteer coordinator, had to deal directly with 1,000 people every day) shared his recollections of this room with me. "It was a designated no-bustle zone," he recalled wistfully. "So it was nice and quiet in there. And only keepers and staff were allowed. Every Tom, Dick, and Harry couldn't go in there, which made it nice for me, because I could go in there and no one could hound me or track me down. I remember sitting down a couple times on the sand, just chilling and watching the chicks, and that was nice for me. It was a haven. And we sat around and laughed, which was great. For me, that was key. If you could keep people's spirits up and keep them laughing–even laughing at themselves–it really made a difference. People can go a whole lot further laughing than they can sort of grumping along."

Everyone who spoke about this room referred to it as "The Sanctuary." This section–replete with the special smell of penguin chicks–was the one area in that massive building that was an oasis of calm and relative order, and it also contained tangible hope for the future of the species. Here, there was new, untainted life. There were no oiled penguins, or sick and emaciated adults. Just healthy, fluffy penguin chicks at the start of their lives–lives that had nearly been snuffed out before they started. But, thanks to their rescuers out on the islands, and to Lauren and Steve's tender loving care, those lives now carried within them the possibility of future generations.

The intensity of the work, the sad physical state of the penguins, our interactions with the wonderful volunteers, and our relationships with our team leaders and fellow team members made for both a challenging and a rewarding experience. Some of us had been working with penguins for just a few years, others for ten or fifteen years–yet egos and self-importance had no place at the rescue centers; neither did thorny personalities or petty grievances. The focus had to be completely on the penguins, and what needed to be done to save them. Although the eight of us had started off as relative strangers, a deep bond rapidly developed between the members of our team, a common occurrence among disaster response workers. It was instantaneous and mysterious. Many of us had not known each other prior to boarding the plane in New York to head across the Atlantic to Cape Town, yet we instantly knew each other's hearts and souls. The essence of our mission, and the reason we all were there, was the same. Each of us cared deeply about these penguins and wanted to ensure they had a future.

South Africa's only penguins had long been struggling to survive and now they were in serious trouble; this oil spill could very well be the event that doomed them to an early extinction. Our goal in coming to Cape Town was to try to save these seabirds, not only as individual animals but as a species. I believe it was because of this important shared purpose that I–for one–quickly came to care very deeply for every member of our team. They instantly had my respect, admiration, and affection. The feelings I have for them still are profound and lasting–and quite unlike the feelings I have for anyone else in my life. We shared a unique, intense, once-in-a-lifetime (hopefully) experience. One does not go through an experience like this unchanged. And one never forgets the people one shared it with. This bond helped us to cope with the difficulties we faced each day in South Africa, and was a source of strength and hope in a seemingly hopeless situation.

The feelings we had for our fellow team members and team leaders extended to the many volunteers we worked with as well. Certain individuals stood out: people who were truly indispensable in helping to keep things running smoothly and efficiently at the rescue centers. These volunteers worked just as hard as we did, and they quite literally gave their blood, sweat, and tears to save the penguins from certain death. But they did more than just help the birds. Though I'm sure they did not realize it, they also helped keep us going. As members of the Oiled Wildlife Rescue Team, we were stressed, burned out, and beyond exhausted; but they cheered us and encouraged us, and did everything they could to help us provide the best possible care for all of the ailing penguins. They lifted our spirits by telling us what a great job we were doing and by bringing us small gifts; and they thanked us over and over for coming to South Africa to help save their cherished penguins. They generously and unreservedly gave the gifts of their time, their goodwill, their compassion, and their friendship.

But the greatest gift these volunteers gave was something less tangible, yet more profound and moving. Just by virtue of showing up and so selflessly being there for the penguins in their time of need, they reminded us that, when given the opportunity, people will do the right thing. This animal crisis truly brought out the best in everyone. We could see, in front of us, individuals from various socioeconomic backgrounds, nationalities, and races coming together to work side by side. For a country that had been divided for so long by apartheid, it was heartening to see people put any differences aside and work together for a common purpose. This was something that struck my teammate Martin as well. "It made me realize that anything is possible," he said years later. "So many people from around the world helped in the effort. White and black people helped alongside each other–an unlikely sight at that time in South Africa's history. And everyone gave their all for the birds, setting aside everything else until the job was done. People would show up at Salt River in white shirts and ties, straight from the office, to feed their beloved penguins . . . people really did care." The people of Cape Town, and from around the world, gave us a strong sense of hope, for mankind and for the animals. While it was easy to be angry at the shipping companies and the other people responsible for what had happened to the penguins, the thousands of volunteers in front of us were living testimony that there was truly good in the world. Ultimately, these 12,500 volunteers restored our faith in mankind.

11

Angels in Oilskins—12,500 Dedicated Volunteers

The volunteers, as you can see, are the backbone of this operation. We would never be able to pay for this operation if it all had to be on paid staff. It's absolutely crucial that we have volunteers, and I salute all of those wonderful people who have come forward to volunteer to help save the African penguins in their darkest hour.

—DR. IAN MACDONALD, WWF-SA

A miracle was slowly unfolding in Cape Town. Thousands of penguins were arriving daily at the rehabilitation centers, and the inescapable question loomed: How would it be humanly possible to provide care for so many oiled birds? Of all the penguin caretakers worldwide, just over a hundred were available to come to South Africa in alternating shifts over the course of the rescue effort, which eventually stretched to three months. Despite the considerable lack of experienced handlers, each of the 19,000 oiled penguins had to be fed, washed, and meticulously cared for. But that care required approximately 1,000 people each and every day. Where would those 1,000 people come from, day after day?

Surprisingly, volunteers began pouring in from all corners of the earth. During those three months a small army of caring and committed individuals flew to South Africa to save the penguins from certain death; some were penguin caretakers or wildlife rescuers, but most were just everyday folks who felt compelled to help. The penguin specialists arrived from zoos and aquariums as far afield as Brazil, England, Ireland, New Zealand, Australia, France, Germany, Poland, Singapore, Canada, Hungary, the Netherlands, and the United States. In addition to these professional caretakers, approximately 12,500 volunteers–none of whom had any experience working with penguins–showed up at the rescue centers. Most of them were from Cape Town, but many had traveled to South Africa from other countries on their own dime when they heard about the oil spill, simply because they cared about these helpless creatures and wanted to do something to help relieve their suffering. By the end of the rescue effort, these indefatigable volunteers had donated more than half a million hours of their time to save the stricken penguins. This unprecedented global response was truly inspiring, and it is the only reason the animals survived their traumatic ordeal. But, in the early days of the oil spill rescue, all of the work was being carried out by just a handful of overworked and overwhelmed rescue workers and volunteers from Cape Town.

To get the word out about the desperate need for volunteers, local radio stations continually broadcast news of the oil spill and the ensuing rescue effort. Newscasters gave daily updates on the status of the oiled penguins and made regular appeals to the general public for help. In an outpouring of compassion and generosity of spirit, the people of Cape Town turned out in droves. And thanks to the efforts of Jeremy Mansfield, an animal lover who is a disc jockey at a radio station in Johannesburg, residents of this city some 1,000 miles northeast of Cape Town were made aware of the need for volunteers as well. Six hundred of his listeners responded by heading to Cape Town; some took the two-hour flight, but most of them climbed into their cars and spent the next fifteen hours driving to the Salt River rescue center. None of these kindhearted people had a real sense of what was awaiting them there. Even after hearing about the rescue effort on the radio, nothing could truly prepare anyone for the shocking scene inside that massive warehouse.

A call center, staffed by just six people, was established to field all of the phone calls coming in from potential volunteers. In the early days of the effort, they received more than 1,000 calls a day; in July alone, these six people answered 28,346 phone calls from local citizens wanting to volunteer their time. From 8 a.m. until 9 p.m. each day, they spoke with people as young as six and as old as ninety-six who phoned in offering to help. Even if they could not do the physical work of washing oiled penguins or cleaning soiled holding pens, people wanted to contribute in whatever way they could; so they donated food, towels, toothbrushes, newspapers, money, and anything else that was needed. One little girl was so worried about the penguins that she insisted her parents drive her to the Salt River Penguin Crisis Centre. Though she wanted to work directly with the penguins, she was much too young to volunteer, so she decided she would donate the entire contents of her penny jar to the rescue effort. This child, maybe seven or eight years old, had just 13 rand to her name (the equivalent of $1.55), but she wanted it all to go to the penguins. Upon arriving at the warehouse, the young girl and her family were escorted directly to the chick-rearing room (a wise decision, given how traumatized many grown-ups were at seeing hundreds of pens filled with oiled penguins), where she gallantly surrendered her life's savings to our teammate Lauren DuBois.

One fortunate thing–if there can be a fortunate thing–about the timing of the oil spill was that it occurred at the start of the South African school holidays. This meant that thousands of energetic and eager students were available for three solid weeks to assist with the effort. Due to the sometimes dangerous nature of the work, a minimum age requirement of sixteen was established for the volunteers; however, this did not keep many teenagers under sixteen from attempting to fly under our radar. More than once, an underage volunteer was discovered and, regrettably, would be relieved of duties and sent home. But during the first chaotic days of the rescue effort, before organized systems had been put into place, help was accepted from whoever wanted to give it, regardless of age. One couple arrived with their two young sons in tow, asking how they could help. As their boys were under ten years old, they were all given the task of cleaning the syringes that were used to give activated charcoal and fluids to the penguins. Until the rescue directors from IFAW and IBRRC arrived, and it was determined that all volunteers must be at least sixteen years old, this family of four sat huddled together around tubs of hot soapy water, happily disinfecting syringes all day.

One remarkable young teenager, Didi Ettisch, was immersed in the effort right from the start. She initially volunteered as part of the construction team, setting up the rooms, holding pens, and swimming pools at Salt River, and preparing the filthy and dilapidated building for the arrival of the penguins. Although she was only fourteen, she worked the same punishing hours as everyone else, typically getting home at 4 a.m., sleeping for an hour or two, and then heading back to the rescue center again by six. Once Salt River was up and running, she was trained how to feed and care for the penguins, and quickly impressed everyone with her competence and reliability. Her work ethic was outstanding, and she handled the duties she was given so well that she was soon designated as a section leader. As such, she was responsible for managing the logistics and manpower for several pools of penguins in a particular area. In this role, she proved industrious, conscientious, and mature beyond her years. Didi skillfully coordinated the care of the penguins in her section, making sure they were all fed and swum every day, in addition to training newly arriving volunteers how to properly feed and handle the oiled birds. And because of her excellent academic standing, once the school holiday ended, her parents and school both granted her permission to skip her classes so that she could continue volunteering at Salt River.

Didi was such an indispensable member of the team that her true age was not revealed when the rescue directors from IFAW and IBRRC arrived and set the volunteer age limit at sixteen. It was only when she collapsed from dehydration one morning a few weeks into the rescue effort that they learned she was just fourteen years old. Despite the fact that she had expertly managed her section for weeks, she was informed she could no longer volunteer and was reluctantly asked to leave. She had worked harder than most people could ever imagine working in their lifetime. But the rescue directors could not afford to start making exceptions to their rules; it would leave them open to too many potential problems. Once her age was discovered, she was not even allowed to go back inside Salt River, so she never had a chance to say her goodbyes to the people who had become like a second family to her.

According to Big Mike, Didi handled the deeply disappointing news with remarkable aplomb and maturity, but she was devastated that she could not see the effort through to the end. The biggest heartbreak for her was not being able to attend any of the releases of the cleaned and rehabilitated penguins. In fact, on the very day she collapsed she had been scheduled to go on her first release, but could not go after her age was revealed. From the first grueling days of the rescue, she had dreamed of seeing the birds returned to the sea. Now, just hours away from experiencing the relief of witnessing the penguins she had cared for set free, this highly anticipated moment was denied her. It was a crushing blow. Although there were other releases after that, she was not able to go to any of them. She is haunted by the fact that she missed out on the experience of finally watching clean, healthy penguins waddle down to the ocean's edge and plunge back into their watery home for the first time in months. Never having seen the positive result of all her backbreaking work, Didi has been left without a true sense of closure. To this day, it still pains her to think about it.

Though a large number of the volunteers in the first few weeks of the rescue effort were teenagers and young adults on school vacation, the volunteer workforce extended far beyond these students. People of all ages and from all walks of life showed up at the two rescue centers every day, ready to work in filthy, stressful, and punishing conditions. There were even several tourists who had flown in to Cape Town from other countries to enjoy a relaxing vacation, but, upon learning of the oil spill, abandoned their plans and spent the rest of their holiday force-feeding penguins in the unprotected outdoor pens at SANCCOB or working inside the dark, putrid-smelling Salt River warehouse getting bitten and poo'd on by ungrateful penguins. Aside from the logistics, penguin expertise, and management provided by the rescue teams, some of the most fundamental aspects of ensuring a successful outcome–feeding and cleaning the oiled birds–fell to these unpaid workers. Without their hard work and dedication over the course of several months, the penguins would not have survived the horror of the *Treasure* oil spill.

There are several challenges to be overcome, however, when relying almost entirely on a volunteer workforce–the first of which is getting each person properly trained and up to speed in a timely manner. The other significant obstacles are retention and consistency. With unpaid helpers, you never know if and when they will show up, so it's impossible to plan ahead or make a schedule based upon how many workers you'll have at any given time on any given day. This requires supervisors to be both flexible and well organized; and it means they must constantly modify their strategy based upon the size and experience level of an ever-changing workforce. On some days at Salt River, fewer than 100 volunteers came to help, while on other days, nearly 1,000 arrived hoping to be of service. But no matter how many people showed up, the same amount of work still had to be done every single day: 16,000 penguins had to be fed and swum; several tons of fish had to be thawed and medicated, hundreds of filthy holding pens had to be cleaned, hundreds of penguins had to be washed, and much more. On days when the volunteer turnout was abysmally low, the work of 1,000 people was carried out by just 200 or 300 people. These were backbreaking days for the intrepid volunteers who had come to help. On those blessed days when the rescue center was teeming with hordes of eager volunteers, the individual workload was lightened, but someone still had to manage the allocation of all those people streaming through the warehouse door. Coordinating the volunteer assignments each day was a full-time job in itself, and an exceptional individual was required to handle these complicated logistics. Luckily, just the right person–himself a volunteer–happened to show up.

Every small army needs a commander–and we had an incredibly

capable and tireless leader in Mike Herbig. "Big Mike," as he is affectionately called, is undeniably the largest human being I have ever stood next to; at six feet ten and 270 pounds, his body is pure, solid muscle. A world-class Jiu-Jitsu champion from Cape Town, Mike was the volunteer coordinator at Salt River, toiling every single day for the duration of the rescue. Having started off as a civilian volunteer himself, he was pressed into permanent service by the rescue directors after they saw how brilliantly he was managing the volunteer workforce. Mike had no prior experience in this capacity, but when he arrived at Salt River a few days after the oil spill and saw a need for someone to direct the volunteers, he stepped in and just started doing what comes naturally to him. He immediately took charge, bringing some order to the initial chaos, and quickly became an invaluable member of the team. But Mike had his own business to manage, so, three weeks into the rescue, he hesitantly informed Jay and Sarah that he had to leave to run his martial arts school. They knew that losing Mike would be a tremendous setback for the operation, and they pleaded with him not to go. Thankfully, they were able to hire him as an IFAW consultant, so he stayed on in an official capacity until the end of the rescue as the Treasure Oil Spill Volunteer Coordinator.

While we were busy wrangling thousands of penguins, Mike was busy wrangling thousands of volunteers. Stationed just inside the entrance of Salt River, he constantly barked orders to the throngs of people lined up to get their assignments, his booming bass voice resonating through the building. And when Big Mike hollered or gave an order, people listened and obeyed. Despite his intimidating appearance, though, he had a warm and gregarious nature. Big Mike was a one-man coach and cheerleading squad; and when time permitted, he made the rounds of the building, praising us for our good work or wrapping us in a huge bear hug. Beaming one of his impossibly wide smiles, he lifted our spirits if we were discouraged or exhausted, always seeming to sense when we needed encouragement most. He was perfectly suited to organize and motivate the thousands of volunteers under his command, and despite his own constant stress and fatigue, always managed to bring a bright spot to our long and arduous days.

One of Mike's favorite expressions, always said with great gusto, was: "You guys *rock!*" Each time he bellowed these words, I was infused with a dose of much needed energy. He instinctively knew (probably from teaching martial arts to so many students over the years) that a little praise goes a long way, especially when someone feels too worn out to continue. Walking into that huge warehouse each day, packed wall to wall with oiled penguins, and realizing the scope of the work that still lay ahead, it would have been easy to get overwhelmed and defeated by the sheer size of it all. But Big Mike wasn't about to let that happen–not on *his* watch.

At the time of our arrival just one week into the rescue effort, things were fairly hectic and confusing. The operation was still growing and the logistics were being fine-tuned. Hundreds of boxes of oiled penguins continued to stream through the warehouse entrance on a daily basis, and most of the volunteers were still learning the ropes. Salt River had only been up and running for a few days, and training hundreds of volunteers each day while trying to coordinate the care of all the penguins in our enormous room was extremely challenging. Once we had devised a tracking system using the dry-erase board, however, it was much easier to manage the care of the 4,500 penguins under our care. Stopping by our room one morning a few days after we had arrived and seeing the system we had come up with, Mike began showering us with praise. "Brilliant!" he roared. "You gals are amazing! I'm going to get dry-erase boards to all the other room supervisors, and show them the tracking method you're using in here. This is *awesome*!" As we were pretty much flying by the seat of our pants, it boosted our morale to know we had figured out a way to make some sense of the monstrous daily challenge of caring for so many animals.

Big Mike's giant voice became familiar not only to those of us confined to the Salt River warehouse but to the residents of Cape Town as well. He was enlisted to give twice-daily radio updates on the rescue operation, and his deep, resonant voice was soon recognizable to thousands of listeners as it traveled across the airwaves, bringing news about the rescue to local citizens. Along with daily statistics, such as how many oiled penguins had arrived at Salt River or how many penguins had been washed that day, he shared lighter bits of trivia, such as how many cups of coffee the Red Cross had served at the rescue center that morning or how many toothbrushes had been used so far to clean the oiled penguins. But, most essentially, in his daily radio spots, he stressed the critical need for more volunteers, urging every able-bodied person within listening range to come down to Salt River to help care for the penguins. In the months following the *Treasure* oil spill, Mike had many encounters, often in unexpected settings, with people who recognized him just by the sound of his voice. About three months after the doors of Salt River closed for the final time, he was at a computer expo with a friend when a woman walking in front of him suddenly spun around and exclaimed, "You're Big Mike!"

"Ya. How did you know?"

"I recognize your voice from the radio," she said. "I heard it every day."

Over and over, the people of Cape Town thanked us for coming to their country to help save their penguins. Their disbelief that we would fly all the way to South Africa from halfway around the world, and their genuine appreciation for our assistance, were palpable; yet we, at the same time, felt tremendously grateful to all of them for their unwavering efforts. After all, we were professional caretakers in each of our home countries, so it was only natural for us to be there during this disaster. They, on the other hand, were students, housewives, manual laborers, business professionals, retirees, the unemployed, even some celebrities–and none of them had any penguin experience prior to volunteering to help these birds. Witnessing so many untrained people willing to endure painful injuries and constant, bone-numbing fatigue all in the name of the penguins was surely as inspiring to us as our presence was to them.

Every one of us there in an official capacity was deeply moved by the incredible volunteer response. In speaking with my colleagues at the time of the spill, and in the years since, it is this aspect of the rescue that always evokes the strongest emotions; most of us cannot talk about the volunteers for long before we are overcome with awe and gratitude.

These tireless volunteers carried on despite punishing conditions and vicious bites inflicted by the wild penguins. These bites soon resulted in my learning a new word. In the United States, when a person is hurt or injured, they usually shout, "Ow!" or "Ouch!" (often followed by a string of expletives). In the Afrikaans language, when someone is hurt, they shriek, "*Ai!*" (pronounced *Eye-eee,* with the emphasis on the high-pitched *eee*). With the penguins' razor-sharp beaks frequently inflicting bloody bites to sensitive parts of the volunteers' bodies, shouts of "*Ai!*" became commonplace, punctuating the relative quiet of the building.

One day, a particularly loud "*AI!*" came from the other side of our room. I rushed over to discover that a young woman had been bitten clear through her upper lip by the penguin she was attempting to feed. This was one of the very real hazards of handling wild penguins, and despite our best efforts to avoid this type of injury by training everyone how to properly restrain these feisty birds, several volunteers were bitten on the face. I sent the young woman out to the first-aid station, where they stitched up her lip and gave her a tetanus shot. The penguin had missed her eye by mere inches, and I had not expected her to return after this incident; but when I looked up again she was back inside a holding pen with a freshly sutured lip, bravely catching and feeding penguins as though nothing had happened. My throat tightening with emotion, I watched her in stunned silence, amazed by her bravery and determination to forge ahead despite having just endured what must have been a very frightening and painful injury.

At one point, in need of more volunteers, calls were made to homeless shelters in the area, in hopes that the offer of a free meal in exchange for several hours of work might entice some folks to come help. Although some of these people initially showed up for the food, after seeing what was happening at Salt River, most ended up staying and working just as hard as–if not harder than–the other volunteers. For many of them, it was the first chance they'd had to feel useful and valuable in quite some time. Many of these people were accustomed to being looked down upon or ignored by others, but being part of this important rescue mission provided an opportunity to feel needed, and to do something that would make a real difference for the animals. One homeless woman in particular left an indelible impression on Big Mike. She was an older woman, and after helping out for two weeks, she came up to Mike one day and gave him a bag of *naartjies* (a South African fruit similar to tangerines) along with an enormous heartfelt hug. "These naartjies are for you," she said, "because you've given us a sense of purpose in what we're doing. And you're always smiling."

He knew this woman was out of work and living in a shelter with absolutely nothing to her name; still, she had gone out and bought him a gift that he knew she could not afford. He was deeply moved by this simple, yet profound gesture. Looking into her eyes, and seeing the gratitude there, this giant of a man was reduced to a puddle of tears. Recalling this moment later, Mike said: "She'd gone and bought me these naartjies because she'd felt useful and needed in that whole context. And that always really struck me, because these homeless folks would do anything you'd ask them to do, no matter how awful it was. They were prepared to do whatever it took, and it gave them a sense of value. Even in the midst of this disaster, along with the bad came some really awesome good–these people, who were society outcasts, felt needed and wanted and helpful and valued." Even to this day, when thinking about this woman or sharing the story about her, he is moved to tears. "It blows me away that it's ten years later, and when I talk about this woman, I still get choked up. I always think of her when people ask about the oil spill," Mike continued. "I wouldn't recognize her if I passed her on the street, and I don't even know if she's still alive–but that, for me, was a really poignant, emotional moment." She was a nameless, faceless, homeless woman in a crowd of more than 12,000, yet she touched him more deeply than anyone else he encountered during the entire three months of the rescue.

On the other end of the social spectrum were the wealthy housewives from the affluent suburbs of Cape Town who came to help. While it was admirable that they wanted to volunteer to help save the penguins, many of these rather well-manicured women did not wish to get their hands dirty or just wanted the "glamorous" job of feeding the penguins. Some of these women showed up in high style, wearing heels and fancy, designer label clothes; clearly, they did not fully comprehend what working with oil-covered penguins in a building filled with coal dust entailed. Some were assigned "clean" tasks such as handing out gloves and boots, or making snacks for the volunteers. And, while many people just wanted to feed the birds, Big Mike had a strict training regimen; volunteers had to prove themselves in other areas before getting to feed the penguins. Everyone, no matter who they were, had to work their way up to becoming a penguin feeder.

There was one well-to-do woman, however, who left a different impression on Big Mike. Seeing her walk through the doors to Salt River with her perfectly coiffed hair, beautifully manicured nails, and a gigantic diamond ring on one finger, he immediately thought to himself, "Oh Lord, here comes another one." Though she was dressed down, class oozed out of every pore, and it was clear to Mike that she was extremely well off. He needed someone to help clean the dirty mats that lined the bottoms of the holding pens, and unlike some of her contemporaries, she seemed amenable to doing this more lowly task. Because the mats were perforated, when sprayed with a high-pressure hose, the water ricocheted back and splattered all over the place, along with the penguin poo that covered the mats. Mike led this woman to the mat-cleaning area and showed her how to operate the fireman's hose, then left her to it. A few hours later, he returned to check on her. The woman's impeccably styled hair was wet and bedraggled and she was dripping with water and guano, yet she had a huge grin on her face.

"How're you doing?" Mike asked.

She turned to him with a beatific smile and said, "I–am–having–*the time of my life.*"

Looking at her incredulously, Mike said, "You must be kidding me."

"You know, Mike," she said, "I haven't had this much fun in *years.*" She was positively jubilant. She returned to Salt River several times after that, and always requested to clean the mats. Mike later said of this woman, "What struck me was the complete contrast. I had so many of these high society ladies arrive, saying, 'I just want to feed the penguins.' But not this woman. She was perfectly happy scrubbing dirty penguin mats all day. She probably went back to her highfalutin home somewhere in a wealthy suburb and entertained her high society friends, but on my watch she cleaned penguin mats, and man, she *loved* it!"

The volunteers were incredibly eager to help, and they sometimes came up with creative and unusual ways of trying to improve things for the penguins. Our teammate Jill Cox had an amusing encounter with one such well-meaning volunteer in her room one day. This woman had flown all the way to Cape Town from Australia, having paid her own way to help with the rescue effort. She had previously worked as a volunteer after an oil spill in Australia, helping to rehabilitate oiled Little Blue penguins. Little Blue penguins are more lightly feathered than African penguins, and as the smallest of the eighteen species they have less body fat, so–even after exiting the ocean and remaining on shore–they are more prone to becoming hypothermic when covered with oil. Once packed tightly into the holding pens inside the warm rescue center, the African penguins we were caring for were at little risk of becoming hypothermic, even with oil still coating their bodies.

Because the pint-sized Little Blues are so vulnerable to the cold, during the oil spill rescue in Australia, the rehabilitation hospital at the Phillip Island Nature Park in Victoria put out a request for hand-knit penguin sweaters to help keep the tiny birds warm while they waited to have the oil washed from their bodies. The sweaters would also prevent the penguins from preening their feathers and ingesting the toxic oil. A sweater pattern was made available to the public, and before long, the penguin hospital was inundated with thousands of tiny, brightly colored sweaters. Some knitters got very creative, making bold striped patterns or using the colors of their favorite sports teams. Once the story hit the internet, the hospital was deluged with even more sweaters made by thousands of concerned knitters from around the globe. It received so many, in fact, that the center had to get two large cargo containers to store all the extra sweaters. To this day, the tiny garments keep arriving. They've started putting the surplus sweaters on stuffed toy penguins that are sold in the gift shop at the Phillip Island Nature Center, and they send some to other penguin rehabilitation centers around the country.

After arriving at Salt River one day, Jill began her usual morning pen inspection to check for any ailing or dead penguins. As she scanned the room, she spotted something in one of the pools that seemed out of place. Walking over to get a better look, she shook her head in disbelief. It took a moment to fully register what she was seeing; but there, implausibly, in the middle of the pool, was a penguin wearing a brightly colored, hand-knit miniature sweater! As Jill stood there, dumbfounded by this curious spectacle, the volunteer from Australia approached and offered an explanation. She had decided this penguin was cold, and thought it could benefit from one of the handmade sweaters she had brought with her from Australia. An African penguin is three or four times the size of a Little Blue penguin (they weigh 8–10 pounds vs. the 2–3 pounds of a Little Blue), so the petite sweater was stretched tightly over the much larger African penguin, making the bird look like a teenaged girl wearing a midriff-baring shirt. Jill had to choke back her laughter–she didn't want to seem unappreciative, but the penguin looked so absurd, and she could just picture the struggle that must have ensued when the woman tried to wrestle the tiny sweater onto this large, feisty penguin. Not wanting to insult the volunteer, Jill thanked her very much and moved the penguin to another pen, where she discreetly removed the skintight sweater.

For those of us working as rehabilitation managers and supervisors, our day lasted from 7 a.m. until 11 p.m. or later. For the volunteers, shifts were broken down into five-hour increments: 7 a.m. to noon; noon to 5 p.m.; and 5 p.m. to 10 p.m. Despite this schedule, many volunteers put in as many hours as we did, working multiple shifts back-to-back, showing up seven days a week to clean the holding pens, feed the penguins, or scrub the thick oil from their bodies. We had several such volunteers in our room who truly stood out, and who eventually became assistant supervisors by proxy. Dennis, Isabella, Erwin, Joy, and Carrots were five of these exceptional individuals, without whose help and unflagging commitment we would have been lost. Their skilled assistance was truly invaluable. Once they were well trained in the handling and feeding techniques, they helped us train and monitor new volunteers as they arrived in our room. There were others who also served in this unofficial capacity, but their names have been lost due to the passage of time and the failings of memory.

Every morning, I was relieved to see Dennis, in his signature white T-shirt (which never stayed white for long), and Isabella, wearing her baseball cap, come into our room, because I knew they would be a tremendous help in getting us through yet another brutal day. And every evening at Salt River, I was buoyed by Erwin's arrival. I never asked Erwin his age but, judging by his salt-and-pepper hair and beard, I imagine he was well over fifty; yet his unflagging energy and limitless endurance were those of a much younger man. Despite the stressful conditions and the sad state of the penguins, Erwin wore a constant sunny smile. Every weekday at 5 p.m. on the dot, he strode into our room to work the evening shift, still wearing his business clothes under his oilskins. Without a word or need for any direction from us, he'd prop his worn brown briefcase against the outside of a pool, climb into the enclosure, and start force-feeding the birds. He wouldn't leave that pen until the very last penguin had been fed. Once that pool was emptied of all the penguins, he'd climb out, pick up his briefcase, set it against the outside of the next pool, and climb in to feed those birds. He'd repeat this ritual again and again until ten or later every night. And the next night he'd return to do it all again.

Not only did Erwin appear every evening after he got out of work, he showed up on the weekends as well–never taking a break from the exhausting work of forcing fish down the throats of thousands of unappreciative penguins. It seemed to be his personal mission to feed as many penguins as was humanly possible. Each of our unofficial volunteer assistants approached their work with the same drive and commitment. Though I tried to persuade Erwin, Dennis, Isabella, and some of the other remarkable volunteers we saw in our room every day to take a day off and get some rest, they were steadfast and determined, and they simply refused to stay away. These amazing people lifted our spirits daily with their dedication, their kind words, and their deep concern for the animals.

Because of its unprecedented size, the oil spill and ensuing rescue effort had attracted international media attention. A week into our stay, a film crew went from room to room separately interviewing each member of our immediate team. During my interview, when asked what I thought about all the volunteers, I started to answer, saying how amazing they were and how grateful we all were to them, but quickly get so choked up that I could no longer speak. Tears sprung to my eyes and time stretched as I silently fought to regain my composure. But before I could do so, the interviewer started weeping, as did the cameraman and soundman, and filming had to be halted while we all pulled ourselves together.

Breaking the unspoken professional code of journalism, the young woman conducting the interview asked if I could use a hug, which I gratefully accepted. News crews and film crews are trained to be objective and reserved, but the scene in that warehouse was devastating, even to them. Like us, though, they could sense the human miracle that was occurring there: thousands of people from Cape Town and from all around the globe had come together, and were working beyond the point of exhaustion, in horrendous conditions, to save the oiled penguins. I apologized to the film crew for crying, but they waved off my apology, stating that the same thing had happened with every other member of our team (male and female alike) and that they were becoming accustomed to the emotional response this question evoked each time it was asked. I am not someone who cries readily, yet, even to this day, when talking about the oil spill, I am sometimes caught off guard when–without warning–my voice catches in my throat and tears blur my vision as I recall the incredible volunteers we worked with each day.

Not only did this rescue effort involve the largest number of live animals ever rescued and rehabilitated, it had the largest volunteer workforce ever assembled in response to an animal crisis; more than 12,500 volunteers had spent over 556,000 hours saving South Africa's penguins from this devastating environmental disaster. Broken down into eight-hour days, this is the equivalent of one person working for 190 years, or 69,500 straight days. A total of 111,200 five-hour shifts were worked; on average, each person volunteered for nine shifts, totaling forty-five hours, though numerous volunteers devoted much more time than this to the rescue effort. Many spent several hundred hours squatting inside crowded, guano-encrusted holding pens force-feeding defiant penguins, or standing over tubs of hot, soapy water with a toothbrush in hand, carefully scrubbing oil off thrashing birds.

While professional rescue staff from SANCCOB, IFAW, IBRRC, and other local conservation groups worked every day of the three-month rescue effort, there were several volunteers who also worked almost daily. They were not paid to be there, and certainly not required to work so diligently, but they did so because they cared deeply about these animals and wanted to do everything they could to ensure their survival. Without all of these dedicated, hardworking volunteers, the penguins would not have survived, and this vulnerable species would be one step closer to extinction. This incredible international response to animals in need shines as an example of the inherent goodness in people, and of the tremendous obstacles that can be overcome when people unite for a common, altruistic cause.

12

Toothbrushes and Dishwashing Liquid—Cleaning the Oiled Penguins

It's just unbelievable. It has been the most incredible effort by our team. These birds were among the worst oiled we've ever seen and, in some cases, have needed up to thirteen washes each.

—KEN BREWER, IBRRC

Unlike many of their avian relatives, penguins are completely defenseless against oil spills. Though any marine bird can have the misfortune of encountering an oil slick at sea, penguins are at a distinct disadvantage because, as flightless birds, they cannot evade this hazard by flying over it. And though they can hold their breath for a few minutes, it's not nearly long enough to enable them to avoid an oil slick by swimming all the way underneath it. Eventually, they have to come up for air, and if they are swimming and hunting in the vicinity of an oil spill, they might surface to breathe right in the midst of a vast expanse of the sticky, black stuff. Penguins—no matter what species they are or where they live—seem to have a disturbing propensity for getting themselves oiled. In fact, penguins have been harmed by oil spills more than any other seabird species in the world—and African penguins more than any other penguin species per capita.

Some scientists have theorized that oil spills actually draw penguins (and other diving seabirds) like magnets, because fish—which routinely seek shelter under objects floating on the surface of the ocean—school beneath the huge, drifting slicks. Hunting penguins are attracted by the schools of fish, which unintentionally set a deadly trap for their predators by luring them to the one thing that will certainly kill them. The only possible way for penguins to survive after being oiled is if they are rescued and scrubbed clean. This was the most critical step in ensuring the survival of the 19,000 oiled penguins at Salt River and SANCCOB: getting them all thoroughly cleaned.

Washing the penguins was a complex and time-consuming process that required patience, a great deal of specialized training, and courage (those beaks were dangerous!). The extent of oiling varied from bird to bird, and it took about four days for each new volunteer to become thoroughly versed in the specifics of de-oiling a penguin. Some penguins were thickly encased in glistening black oil from head to toe, looking as though they had been submerged in giant vats of sticky tar. Many more were stained solid brown with a lighter coating of oil, looking as if they had been quickly dipped in milk chocolate. Others still had small splotches of the stuff dappling their bodies.

The most heavily oiled penguins were washed first, as they were at the highest risk of dying from the toxic effects of the oil. The others had to wait their turn; for many of the birds, that wait would be as long as two months. One might assume that the birds were sent directly to the washroom the moment they came through the door, but the process was not that straightforward. With thousands of oiled penguins arriving daily, more birds were coming in than could possibly be processed. Plus, it had been learned through prior experience that washing oil-covered penguins immediately after admission to a rehabilitation center was actually detrimental to their health and well-being. The traumatized animals needed some time to adjust to their new surroundings before going through the rather long and stressful process of having the oil removed from their feathers. This waiting period is known as *stabilization.* This important two-day interval was not yet standard practice when the *Apollo Sea* sank, and was found to be one of the contributing factors to the high mortality rate during that event.

During stabilization, the penguins were mostly left alone to give them time to acclimate and calm down following the stress of being captured by humans, tossed into boxes, trucked to a dark, dusty rescue center, and thrown into a pen tightly packed with a hundred other birds. But first, each penguin had to go through a comprehensive intake process, during which the following steps were taken: Upon admission, each bird was examined and evaluated by the head veterinarian to determine its health status and general condition. After this initial evaluation, a medicated ointment was placed into the penguin's eyes to protect them from the caustic oil; the corneas developed ulcers from contact with the oil, and, if left untreated, the birds would go blind. Most penguins also had a black coating of oil inside their mouths from trying to preen their feathers, so any oil was swabbed from the delicate mucous membranes inside the bird's mouth. After this, the penguin was tube-fed activated charcoal to mitigate the internal damage from oil ingestion.

To administer this thick liquid substance, a large syringe containing the activated charcoal was attached to a flexible rubber catheter. An experienced handler, restraining the penguin between her or his legs, forced its beak wide open, while a second person would carefully thread the long catheter into the bird's esophagus, passing it all the way down into the stomach. This required some training to ensure the tube was not inserted into the trachea, which lies just in front of the esophagus. If the liquid charcoal was accidentally injected into the trachea, the penguin would end up with lungs full of thick fluid, which is not conducive to breathing, to say the least. Once the catheter was safely in place, the plunger on the syringe was pushed to inject the activated charcoal. The toxins in the oil bound to the charcoal, preventing them from being absorbed through the bird's intestinal tract. The toxins were then excreted, along with the activated charcoal, when the penguin defecated. Bleeding ulcers were another harmful side effect of oil ingestion, so the activated charcoal also had kaolin in it, which lined the bird's stomach to protect it from the caustic effects of the oil.

Immediately following this procedure, the penguin was hydrated with a solution called Darrows. This fluid contained a mixture of glucose and electrolytes, and helped to provide some much needed energy for the dehydrated and exhausted birds. The Darrows was administered using the same technique that was used to give the activated charcoal; the same catheter could even be kept in place, and a new syringe containing the Darrows attached to it. Penguins that were very dehydrated or critically ill were given another dose of fluids subcutaneously, in addition to the fluids they received orally. To give these subcutaneous fluids, a needle was placed under the skin on the penguin's back and the fluids directly infused.

Because ingested oil starts to break down red blood cells once it is absorbed into the bloodstream, some blood was drawn and tested to determine the penguin's iron levels, as well as its hydration levels. To combat anemia stemming from the destruction of red blood cells, an iron injection was given in the chest muscle beside the penguin's breastbone (called the *keel*). Ingested oil also interferes with the absorption of certain vitamins, so a B1 injection was given in the chest muscle as well. Finally, the penguin's blood sample was examined for the presence of bloodborne diseases, such as avian malaria and babesiosis. If any of these diseases was detected, the penguin was separated from the rest of the birds and either sent to the ICU at SANCCOB for critical care, or euthanized if it was too severely compromised. Once this intake process was completed, the birds were left alone for about forty-eight hours to chill out.

On July 1, 2000, the washing of the oil-soaked penguins got underway at Salt River. Although there were already more than 15,000 oiled penguins in the building by then, just 59 were washed that day. During the initial days of the rescue effort, staff and volunteers were only able to clean about 200 penguins per day. This was partially because only one washroom had been set up so far, and partly because the volunteers were just learning how to clean the penguins.

The process of training people always slows down the work being done, but it was critical that each volunteer had a thorough understanding of the methodical procedure. If not washed properly, the birds might have to go through the whole washing process a second time, extending the length of the entire rescue effort. Because the intensive training took four days, it was almost a week before the first volunteers were up to speed. In addition, the hot water system in the washroom did not always work, so things would grind to a halt whenever the water ran cold. As anyone who has washed a greasy pan knows, without hot water it's nearly impossible to break down an oily substance. Without a reliable hot water system it would take forever to remove the thick bunker oil from the penguins' feathers.

Frustration and concern mounted as the rescue directors realized how slowly the washing of the penguins was progressing. During our first week at Salt River we often saw Jay, calculator in hand, muttering to himself as he tried to work out just how long it was going to take to wash 19,000 oiled penguins. Throughout each day, he anxiously tapped away at the buttons, doing the math again and again; you could just see him *willing* more agreeable numbers to appear on the small LED screen. But no amount of massaging the numbers could change the facts. After the first three days of washing, when Jay recalculated how long it would take to clean the rest of the birds, the dreadful reality of the projected time frame sunk in. Given the current rate at which the penguins were moving through the system, it would take nearly *six months* to wash and rehabilitate all of them. This simply was not an acceptable time frame.

It became an immediate priority to find a way to get more birds washed every day. This was imperative, as housing thousands of wild penguins indoors in overcrowded and unnatural conditions for that length of time would lead to even more health issues. The stress alone could compromise their immune systems, making them more susceptible to illness and disease. And, while the exact ramifications of being coated with toxic oil for several months were not fully known, it's safe to assume that it would be deleterious to their health. Not only that, but who would help care for the penguins if they remained at Salt River for six months? Finding enough volunteers who were not only available but also willing to help for such a long period of time was highly unlikely. On top of all this, the penguins were eating a tremendous amount of food: up to 10 tons of pilchards per day. Obtaining this much fish in the short term had already presented a challenge, and it was not known if such a large amount of pilchards could be caught and brought to Salt River every day for the next 150 days. And finally, would the rescue team directors be able to sustain the relentless pace of such grueling work every day for six uninterrupted months without driving themselves into the ground from the stress and exhaustion?

Cleaning each penguin was a laborious and painstaking procedure, and only the most reliable volunteers were selected for this duty. If they had washroom volunteers who flaked out after a few days, the whole washing operation would have been slowed down while new volunteers were trained to replace them. Because of the critical need for consistency with this particular task, the rehabilitation managers only chose volunteers they felt certain would show up day after day, even if they were exhausted and didn't really feel like scrubbing penguins some mornings. During the training process, it took an hour or more for two people to wash one mildly oiled bird. Even working at peak efficiency, it still took an average of forty-five minutes for two fully trained people to clean one penguin. Washing the penguins whose bodies were thickly encased with the viscous oil took much longer.

Washing the oil off a wild African penguin is probably a little like trying to hold on to and bathe a soap-covered, aquaphobic toddler who's in the midst of a raging temper tantrum. The same ingredients are all there: soap, water, and a small squirming creature resisting your efforts. Only the penguin hurts a lot more when it bites. The process for each bird required two experienced volunteers, several tubs of hot water, degreaser, dishwashing detergent, a toothbrush, protective clothing, and a high-pressure hose. One heavily gloved volunteer restrained the penguin, while the second volunteer sprayed it with a first-stage degreaser. This degreaser was made from a light vegetable oil, which started the process of breaking down the heavier fuel and bunker oils covering the penguin's body. After marinating in the degreaser for a while, the oil on the penguin's feathers began to soften, and then the real cleaning could begin. The penguin was placed in a washbasin filled with hot soapy water, and while one volunteer restrained the bird, the other one (also wearing thick gloves) meticulously scrubbed every inch of the penguin using their fingers. For more stubborn spots–as well as delicate areas such as the face–a toothbrush was used. The heavy rubber gloves were not just for protecting everyone's fingers from the birds' sharp beaks–they also helped to protect handlers from the toxic oil (which might be harmful to humans as well if absorbed through the skin). Because the oil had seeped through the penguins' dense feathers, they were carefully separated to clean the downy underlayer, as well as their skin, which was very sensitive to the oil. (Some fuels and oils are so caustic that penguins will suffer chemical burns to their skin, eyes, and mucous membranes. Even their lungs, trachea, and digestive tracts can be subjected to these burns.)

Once the water in the washbasin turned brown and murky, the penguin was transferred to a new tub of hot, soapy water, and the process was repeated. The two volunteers did this for as many times as it took for the water in the basin to run clear, indicating that all of the oil was finally off its feathers. Most penguins required at least three or four thorough washings, and more heavily oiled penguins went through ten or more before the oil was completely removed. Great care was taken to remove all traces of the oil; if a spot the size of a quarter remained, the bird was returned to the washroom to be cleaned again. Even though the rest of its body may have been clean, if a penguin was released with the smallest amount of oil on it, the icy ocean waters would seep through that one compromised spot in the feathers, forcing the bird to seek the warmth of land. If a penguin can't stay warm while at sea, it won't be able to hunt for enough food to sustain itself, so it was vital to remove every last speck of oil. Once the washers were certain there was no remaining oil on the feathers, the penguin was rinsed with fresh water, after which it was put under heat lamps overnight to dry out and warm up again.

Under ideal conditions, after being washed, each penguin would be rinsed with high-pressure hoses for twenty minutes to remove all of the residual soap; however, with 19,000 penguins to clean, this part of the process had to be drastically shortened. If twenty minutes were spent thoroughly rinsing the soap from each penguin, it would have dramatically extended the time to wash all the birds. Not only would this have increased how long each bird would have to sit covered in oil while waiting to be washed, it would have also lengthened their overall stays at the rescue centers. So, the goal was to get the penguins through the system and back out into the wild as quickly as possible.

Therefore, after a brief five-minute rinse, each penguin underwent enforced swimming sessions for a week or two to rinse the remaining soap from its feathers. Every day, the penguins were herded into shallow pools for short swims, which increased in length with each passing day. This process was critical in enabling the birds to waterproof their feathers again after being cleaned. Any residual soap would compromise their ability to waterproof their feathers, so a penguin could not be released until all the soap was gone and its downy undercoat was still dry after swimming. Shortening the rinse cycle was the only possible way to get all of the oiled penguins cleaned in a reasonable amount of time. Had all 19,000 penguins been given the usual twenty-minute rinse after being washed, it would have added another 5,000 hours of labor to the process, and the entire rescue effort would have lasted another few weeks.

Jay's concerns about getting all of the penguins washed more expeditiously were eventually relieved when a second washroom was built, a reliable hot water system was installed, and dozens more volunteers were finally up to speed on the washing procedure. With ten to fifteen stations in each washroom, and with a more experienced volunteer crew, the cleaning began to move along more efficiently. It took about two weeks to reach this stage, but once everything was established and running smoothly, between

400 and 550 birds were washed each day. On their best days, the washing teams were cleaning close to 600 penguins a day. The most birds ever washed in one day was 757; this record-breaking event occurred on July 26, as they were nearing the completion of the washing phase. But it was the very first bird to be cleaned, on July 1, that evoked the strongest emotions for many of the people who were involved in the rescue effort.

This significant moment is etched in Big Mike's memory. "One event that really stood out for me was when we washed the first bird," he recalls, his voice brimming with emotion. "We'd just gotten the whole system going, with boilers, etc. This one bird was washed and rinsed, and then they brought him to the drying room. It was about eleven at night, and we were just shutting down the center, and I remember standing there looking at this flipping penguin that was bloody black and white. All I'd seen for two weeks were solid black penguins. And seeing this incredibly clean, pristine, black and white African penguin standing there–looking a little bit peeved, by the way–I think everything caught up with me. I can remember standing there with tears streaming down my face, thinking, 'This one guy represents success. If nothing else, *this* guy has been saved. *This* guy, he's good, he's cool. He's going to make it.' That penguin, to me, represented a beacon of hope. 'This is why, for the last two weeks, I've been putting in eighteen-hour days,' I thought. *'This.'* It was like a pinnacle result. I remember that as a particularly poignant moment."

One of the first questions people often ask about removing the oil from the birds is, "Did you use Dawn to clean the penguins?" Ironically, just two years before the *Treasure* oil spill, we had hosted a "De-oiling Wildlife" workshop at the New England Aquarium. Staff from Tri-State Bird Rescue and Research had come from Delaware to teach the workshop, during which they informed us that Dawn was the best product for washing oiled wildlife: it was gentle, non-toxic, and very effective at breaking down oils. (Since learning this, it's the only dishwashing soap that's graced my kitchen counter.) And now, Heidi Stout from Tri-State was functioning as the head veterinarian at Salt River. But Dawn was not available to us during the rescue, so another safe and effective detergent was utilized to remove the viscous oil from the penguins' feathers.

A California-based company, GNLD, and its distributors in Africa donated more than $11,000 worth of their light duty concentrate (LDC) to the *Treasure* rehabilitation effort. More than 7,500 liters of the detergent were used to clean all of the oiled penguins. The company donated half of the LDC used during the rescue, and they encouraged their many distributors to donate the rest. GNLD's impassioned appeal to their distributors was: "Buy a liter of LDC and save a penguin." (There is even a video on YouTube about the company's involvement with the *Treasure* rescue, called *GNLD Operation Penguin.*) This household cleaner was both mild enough to be safe for the animals and highly effective at dispersing fats and oils, making it ideal for removing the thick bunker oil covering the penguins' bodies. It was also suitable for cleaning just about everything else–from the Dri-Dek matting to the holding pens to the syringes and catheters used to give fluids and activated charcoal to the penguins.

One of the most remarkable aspects of this cleaning process stars a teenager named Louis Kock, who was a medical student at the University of the Free State in Bloemfontein, 700 miles northeast of Cape Town. It is common practice when de-oiling wildlife to spray the affected animal with a light oil first (typically a vegetable oil), which helps to break down the heavier oil coating their feathers or fur. Different products have been tried over the years: olive oil, maize oil, and others. They often used canola oil at SANCCOB, but they found it left a greasy film that was difficult to remove. About two years before the *Treasure* oil spill, this young student–whose father was a microbiology professor at Free State–had invented a new formula for a highly effective first-stage degreaser using sunflower oil and ethyl alcohol. Created as a science fair project, his original objective was to come up with a safe and inexpensive product to replace the used cooking oils that are sold off-market at very low cost to poor South Africans. Unable to afford to buy new cooking oil, impoverished people commonly buy this used oil, then use it over and over to cook with, even after it has turned jet black, by which time it has become carcinogenic.

After developing his product, Louis realized it could have other applications, and he became interested in seeing if it might be useful as a degreaser for oiled wildlife. After speaking with Estelle van der Merwe at SANCCOB to learn more about cleaning seabirds, he spent a great deal of time experimenting with his formula, testing it on loose bird feathers until he was satisfied with the results. He then gave some to Estelle so she could see how well it worked on live oiled birds. It was important to test the product for safety and efficacy, and she wanted to get an unbiased assessment, so she conducted a covert experiment. Not wanting the staff to be aware that she was testing a new product, she took a heavily oiled gannet–one on whom the oil had literally hardened into a thick shell–and secretly sprayed one side of its body with the young man's degreaser.

After letting the product soak in for a while, she handed the bird off to Charles, her most experienced bird washer, and waited to see what happened. After some time, Charles–looking rather befuddled–approached Estelle, asking her to come see the bird. He and the other staff couldn't understand what was going on. They had washed the bird as usual, but one side of the gannet looked clean and beautiful, while the other side was still stained brown with traces of oil. On the side that Estelle had sprayed with Louis's product, the caked-on oil had softened quickly, making it much easier to remove. With that, Estelle knew they had found their new first-stage degreaser, and asked Louis if he could produce more of the product. He did, and they began using it regularly at SANCCOB. In fact, one year prior to the *Treasure* oil spill, it was used successfully during the rescue of 300 oiled Cape Gannets.

But when the *Treasure* oil spill occurred, and there were suddenly 19,000 oil-covered penguins in need of washing, not enough of the new degreaser was on hand to clean so many animals. At the beginning of the rescue effort, the product was donated to SANCCOB; but as the size and scope of the operation became apparent and more degreaser was needed, SANCCOB began purchasing large quantities of it. Because this degreaser was so effective at quickly breaking down the heavy bunker oil, it greatly reduced the time needed to clean each penguin. Not only did this get the birds through the system faster, but it reduced the stress on each animal as well, because they didn't have to be restrained for quite as long. Partway through the rescue effort, though, they ran out of degreaser. There was literally none left–not at the rescue centers and not at the lab where it had been produced. In a panic, Estelle called Louis and his father, begging them to make some more as quickly as they could. They worked around the clock for the rest of the rescue effort to produce enough degreaser. Thanks to their tireless efforts, enough of this product was available to clean every oiled penguin at Salt River and SANCCOB. I still find it extraordinary and profoundly inspiring that a teenager–a seventeen-year-old kid–invented a product that helped save the lives of 19,000 penguins. How cool is that?

13

Peter, Pamela, and Percy's Long Swim Home—The World Anxiously Watches

It's the biggest dilemma I've had to face in my life. If we hadn't fenced in the island, the birds would have gone to sea and they would have got into the oil. It doesn't matter that the oil isn't on the beach; it's the oil in the feeding grounds that counts.

—DR. TONY WILLIAMS, CNC

During our first day at Salt River, the place had been abuzz with news of a disturbing new development out at sea. It seemed the shifting winds and ocean currents had pushed the drifting oil slick onto the shores of Dassen Island the previous afternoon. Like their counterparts that had already been rescued from Robben Island, the 55,000 penguins that lived on Dassen Island now faced the very real threat of getting oiled as well. About half of the penguins were nesting on the southern end of the island, and the oil had polluted the beach adjacent to their breeding grounds. The rest of the penguins were located on the other end of the island, where it was hoped they would be at lower risk of coming into contact with the drifting oil. A low wall, which had been built around the perimeter of the island near the turn of the century to aid in egg collection, had been repaired, and about 10,000 of the penguins had already been moved to the inside of this wall to prevent them from entering the ocean and getting oiled.

In addition, four miles of fencing had been used to cordon off three main nesting sites located outside this wall to keep the penguins on shore; but a week had passed since they had been fenced in and rescuers were concerned that they would soon start starving to death. Unfortunately, there was neither the staff nor the food available to force-feed the penguins trapped on the island. But if they released the birds to let them eat, they would undoubtedly get oiled as soon as they entered the water. By the time this impending crisis revealed itself, SANCCOB and the Salt River warehouse were filled to bursting with oiled penguins, and the volunteers who had come to help were already stretched beyond their physical and emotional limits. It seemed that the penguins on Dassen Island were either going to die from starvation or they were going to perish because there would be no way to rescue and rehabilitate them once they got oiled.

Faced with this new predicament, penguin researchers, wildlife rescuers, and conservation officials gathered for an emergency meeting to discuss their options. It quickly became clear that there were no straightforward answers or ideal solutions. Because of the variable weather conditions and shifting ocean currents, they had no way of predicting exactly how long it might be before the oil slick moved away from the island. Normally, strong winter winds and choppy waves would start to break up and disperse the oil, but the weather that week had been unseasonably calm, and the oil slick was persisting longer than usual. It may or may not linger in the area. They just did not know. Even if the oil retreated from the coastline of the island, the penguins were still at risk, because part of the slick still floated on the ocean's surface in their feeding grounds. If the penguins were allowed out to sea, they would most certainly come into contact with the oil as they dove beneath it, hunting for fish.

The principal decision makers were faced with an impossible choice. Realizing the resources did not exist to deal with another 55,000 oiled birds–and still not sure how long the oil would remain near Dassen–a radical and untested plan was proposed by Dr. Rob Crawford, a senior conservation official with Marine and Coastal

Management. They would remove as many penguins as they could from the island and transport them up the coast to Cape Recife, in Port Elizabeth, where they would be released into clean waters. It was not feasible to move all of the birds, so they decided to concentrate their efforts on those most likely to get oiled–those closest to the oil slick, on the southern tip of the island. Cape Recife was chosen as the release site because of its proximity to the second largest African penguin breeding island, St. Croix Island, and because of its distance from Dassen Island, which lay some 560 miles to the west, just past Cape Town. There were abundant fishing grounds near St. Croix, and it was hoped that the relocated penguins–who had not eaten for a week now–would spend some time feeding there before embarking on their long swim home.

Researchers knew, from prior releases, that it would take the penguins two to three weeks to swim back to their breeding islands from this location. They did not have a lot of data to go on, but knew of a group of 150 penguins that had been oiled at St. Croix Island in 1979 and brought to SANCCOB for treatment. After being rehabilitated, one of the birds from this group was released on Robben Island, and eleven days later it was back on St. Croix Island, having averaged 2.1 mph on its 560-mile journey. Using this bird as their model, the hope was that, by the time the relocated penguins returned to Dassen Island, the oil would be cleaned up from the ocean and the coastline. But there was no guarantee it would be.

The people charged with making this decision had no way of knowing if their plan would actually work. Dassen Island was home to one third of the entire world population of African penguins, so it was essential to protect as many of them as they could. Wildlife rescuers had never attempted a large-scale relocation like this before. It was essentially a huge experiment–and an extremely risky one at that. Local researchers were assuming that the penguins would be able to navigate their way back to their home islands; although birds are known to have excellent homing instincts, these penguins had never made this journey before. If they removed the penguins, and the oil slick suddenly receded or dispersed, they would have unnecessarily jeopardized the lives of thousands of birds by attempting to relocate them. Not only would their lives have been put at risk, but the chicks they had been raising would be sentenced to certain death. Left behind in the nests as their parents were collected for relocation, the young chicks would be exposed and defenseless. If they did not fall victim to predators, these helpless chicks would eventually succumb to the elements, or they would slowly starve.

Armed with the knowledge that the chicks have only a 15 percent chance of reaching adulthood, rescuers had to focus their initial efforts on capturing the unoiled adults and moving them from harm's way. As heartbreaking as it was to leave the chicks behind, it had to be done for the survival of the species. Once rescuers had removed the adults, they would return to the island to rescue the chicks. There were many unknowns at this juncture, and moving the clean penguins was a calculated risk. But they just could not afford to have any more penguins get oiled at this stage of the rescue effort. At least with this plan, there was a chance these penguins would survive the *Treasure* oil spill.

After making one last flyover on July 1 to assess the path of the oil slick, the researchers knew what they had to do. The oil slick remained stubbornly along the coast of the island, so they would have to start moving the penguins off Dassen the next day to keep them from starving or getting oiled. The massive evacuation was soon under way. They devoted the next three days (July 2–4) to collecting clean penguins for transport to Cape Recife. After removing as many clean birds as they could from Dassen Island, rescue workers collected any oiled adults for rehabilitation and released the remaining penguins to allow them to hunt. They could only hope and pray that these birds would not get oiled as they made their way to their foraging grounds. In addition to the 12,345 clean birds removed from Dassen Island, 7,161 clean birds were collected from Robben Island and brought to Cape Recife as well. Once all the adult penguins had been dealt with, rescuers returned for the abandoned chicks. To their great relief, they discovered that most of the chicks on Dassen Island were on the verge of fledging and did not need to be rescued after all;

close to the age at which their parents would have kicked them out of the nest anyway, they were likely to manage on their own. After a careful search through the nesting sites, 707 younger chicks were rescued and brought to the mainland for hand rearing.

To avoid the same problem of asphyxiation that had occurred during the *Apollo Sea* rescue, this time the penguins were transported in sheep-shipping trucks. These trucks had an open structure, allowing for plenty of fresh, circulated air during the sixteen-hour journey to Cape Recife. Each three-tiered truck carried approximately 2,000 penguins inside cardboard boxes that had been newly designed and built for this exact purpose in the six years between the *Apollo Sea* spill and the *Treasure* spill. The main criterion was that the transport boxes be well ventilated, so each one had thirty-two round holes in it, as well as openings on both ends for carrying. The only drawback was that when the penguins spied human hands entering these openings, they quickly chomped on the intruding fingers.

More than 10,000 transport boxes, flattened and stacked, were shipped out to Dassen Island, where a team of forty volunteers spent their days doing nothing but assembling them. Despite everyone's best efforts after the *Apollo Sea* rescue to prepare for the next big oil spill, they had not anticipated needing so many boxes, and the supply soon ran out. Estelle made an emergency call to Nampak, the company that produced the boxes, to plead for more, and employees at the factory worked straight through many long nights to be sure there were enough to rescue the penguins from the islands. During the *Apollo Sea* rescue, one of the factors contributing to the loss of so many penguins was that too many birds had been put into each transport box. This time, each box (measuring 22.5 inches x 14 inches x 16.5 inches) held just three adult penguins. To expedite the removal of the penguins from Dassen Island, however, the birds were packed five to a box during the initial phase of the evacuation. After being transported to the mainland by boat and helicopter, they were repacked three to a box before being loaded onto the trucks heading to Cape Recife.

The difficult lessons learned during the *Apollo Sea* spill enabled wildlife responders to safely move four times as many penguins after the *Treasure* sank. This time, the refinement of techniques used to transport penguins was a critical factor in reducing the loss of life; still, a small number of the birds (just 241) did not survive their overland journey. Despite everyone's best efforts to ensure there was adequate ventilation, 160 birds succumbed to carbon monoxide poisoning from truck exhaust fumes. Another 81 of the relocated birds died just after release, primarily from starvation, as it had been a week since they had last eaten.

Although it was upsetting to lose any penguins, at least this time the amount lost was a mere fraction of the total number of birds handled throughout the course of the rescue. Of the 19,000 oiled penguins brought to the rescue centers and the 19,506 clean penguins relocated to Cape Recife, just one half of 1 percent (0.5 percent) lost their lives while being transported. Compared to the devastating loss of nearly 5,000 penguins (50 percent of the birds handled) during the transport phase and first few days of the *Apollo Sea* rescue, this can only be seen as a positive result. Indeed, in a paper published by Dr. Rob Crawford and a team of scientists shortly after the *Treasure* oil spill, it was pointed out that, had all of the penguins been left on Dassen Island, and had they all been set free to feed in their oil-contaminated foraging grounds, the death toll would certainly have been far greater.

Because no one had ever tried to relocate so many animals at once before, the researchers weren't exactly sure what would happen. They had to trust their knowledge of African penguin biology and behavior, and hope that these birds did what was expected of them: enter the ocean, turn right, and swim for two to three weeks until reaching Cape Town. It would be a perilous journey; the route was new to them and there would be numerous predators to evade on the 560-mile-long trip. To monitor the progress of the penguins, three individual birds were fitted with satellite tags; two were from Dassen and one was from Robben Island. If they made it safely back to their breeding islands, there was a good chance that the rest of the penguins would survive the long journey as well.

Though the actual genders of the three tagged penguins were unknown, it was presumed from their sizes that two were males and the third was a female. All three wore the same expensive hardware: at a cost of $2,000 each, the satellite tags they carried would provide scientists with important data regarding the African penguins' navigational abilities and swimming speed. Just 3.6 inches long by 1.9 inches wide and 0.8 inches deep, these 4.4-ounce devices were affixed to the penguins' backs using Velcro strips and epoxy resin. One strip of Velcro was attached to the underside of each satellite tag and the other to the feathers on the bird's back. The transmitters would be removed once the penguins were caught following their return to their breeding islands. If they were not caught after they arrived home–as turned out to be the case with the ever elusive Peter–the tags and Velcro strips would fall off in about a month's time as the glue broke down. In addition to the satellite tags, all three penguins were fitted with stainless-steel bands that had individual identification numbers engraved on them. Of the 19,500 unoiled penguins released at Cape Recife, 3,359 were fitted with these ID bands, which were placed at the joint where the penguin's wing meets its body. These bands would allow researchers to identify each penguin in the future. By monitoring the movements and recovery of the birds after the oil spill, scientists would glean important information about the success of the rescue and rehabilitation efforts, in addition to general population statistics.

The time had come to set the penguins free. The press was on hand to record the highly anticipated releases of the satellite-tagged penguins, and the world watched closely as these three–dubbed Peter, Pamela, and Percy–began their long swim home. With them went the hopes and prayers of a nation. Peter, who hailed from Robben Island, was released first, on June 30. The two Dassen Island penguins, Pamela and Percy, were released on July 3 and July 5, respectively. The satellite transmitters they wore (called ST-10 PTTs, or platform transmitter terminals) were funded by SAP Africa (a business consulting firm) and supplied by Telonics, a company based in Arizona. As the pioneering penguins swam toward home, data from their transmitters was periodically beamed to satellites orbiting the earth and downloaded to the Argos Services Centre in Toulouse, France. From there, the originating location of each transmission was forwarded to Professor Les Underhill, a researcher and statistician with the Avian Demography Unit at Cape Town University. He then posted the coordinates to an animated map on the ADU's website, enabling local scientists to monitor the progress of the birds. Following their release, the remaining 19,343 relocated penguins were set free in large batches over the next few days.

But the swim home would not be easy. These penguins were already stressed and disoriented, and now they had to embark on a marathon swim through uncharted territory. They had never navigated their way from this location before, and the waters they were swimming through were some of the most shark-infested in the world: Great Whites and dozens of other shark species inhabit these icy waters off the coast of South Africa. Large numbers of Cape fur seals prowl this coastal region as well, and any one of these formidable predators would gladly eat a penguin if given the opportunity. In addition, the three birds were starting their 560-mile journey in an already weakened state, having lost a great deal of weight after not eating for a week or more. It would be a perilous trip, not only for Peter, Pamela, and Percy, but for all of the penguins trying to find their way back home.

Researchers in Cape Town anxiously followed the satellite transmissions of Peter, Pamela, and Percy as they embarked on their long journey. Each day, the penguins' routes were posted on the ADU's website, and each day, the blips on the monitor showed they were heading in the right direction. Pamela, though, had everyone biting their fingernails for a while: she hung around Cape Recife for a few days before eventually starting to swim west toward Cape Town. Researchers and rescuers watched with increasing hope as each day passed and the penguins kept moving closer to home. They began to believe that this crazy, improvised plan of theirs might work after all. But then came a disturbing report of squid fishermen at Cape Recife shooting at the just-released penguins. Apparently concerned about competition for their catch, they were trying to scare off a large group of the hungry birds. The shooting continued unabated for forty-five minutes until fishermen on neighboring boats informed the perpetrators that these were the penguins that had just been rescued from Dassen Island. Once made aware of this, they halted their lengthy attack and allowed the penguins to continue on their way.

Thousands of people from around the globe visited the ADU's satellite tracking page on a daily basis to follow the progress of the three penguins, who had captured the attention and the hearts of the world.

The ADU's website was inundated as people checked in to see where the penguins were and to cheer them on from afar; during the few weeks that it took the birds to reach their breeding islands, the ADU's satellite tracking map was viewed more than 100,000 times. A collective global sigh of relief was released when, after an eighteen-day trip, Peter arrived on Robben Island on July 18. Percy landed on Dassen Island on July 20 after a fifteen-day jaunt; and Pamela brought up the rear, arriving on Dassen Island on July 25, twenty-two days after leaving Cape Recife. When they saw the satellite transmissions indicating that all three penguins had made it safely home, researchers finally knew they had made the right decision. Their risky plan had worked.

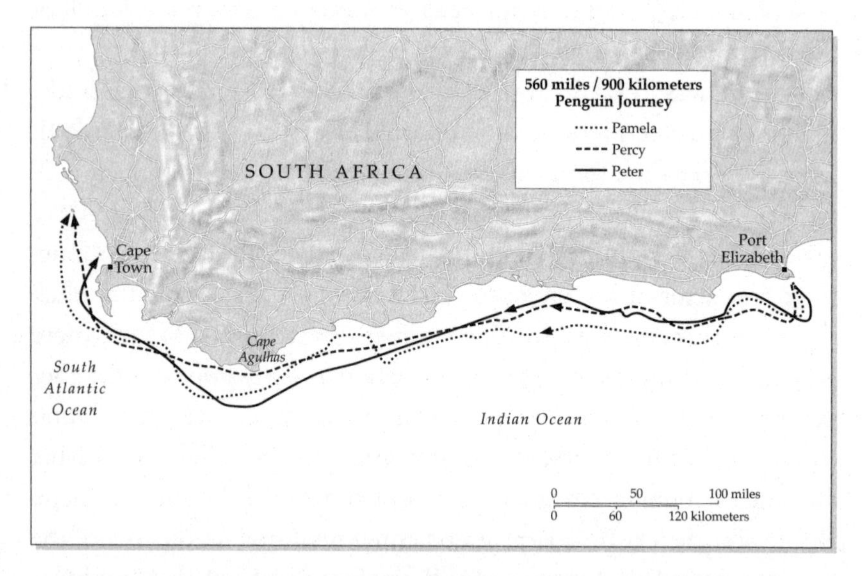

Map showing Peter, Pamela, and Percy's 560-mile swim from Cape Recife, Port Elizabeth, to their islands near Cape Town.

During their grand–albeit involuntary–adventure, the birds had become a media sensation. When they finally reached their respective homes, throngs of reporters descended upon the islands, vying to capture the first photographs of the world's most famous penguins. But the penguins didn't make it easy for the press. Percy was the only one to make an appearance while the reporters were there. The press was invited to a "welcome-home" party for Peter, but poor weather prevented them from going out to the island that week. Because he arrived home first, Peter was recognized by *Time* magazine as one of their Winners of the Week for his gutsy swim back to Robben Island. His celebrated journey was even featured on *The Daily Show,* with Jon Stewart labeling Peter, a Penguin Hero for his remarkable achievement. The media touted the penguins' swim as a daring adventure, rife with threats from predatory sharks and seals, and the fanfare brought even more attention to the huge rescue still under way in Cape Town.

Using these three penguins as their inspiration, a local supermarket came up with a unique fund-raising campaign to help the penguins undergoing rehabilitation at the two rescue centers. Working with Cape Fish Supplies, Pick 'n Pay market developed a new pet food, which they debuted on July 27 at the Salt River Penguin Crisis Centre. It was called "Peter, Pamela and Percy Pure Pilchards Pet Food," and ten cents from every can sold was donated to SANCCOB to help defray the cost of the mammoth rescue effort. The label read: PETER, PAMELA AND PERCY ARE THE THREE PENGUINS REPRESENTATIVE OF THE GROUP OF PENGUINS CLEANED AND FED AT SALT RIVER. A longtime supporter of SANCCOB and of South Africa's penguins, Pick 'n Pay markets in the Cape Town area set up collection centers where people could drop off towels, newspapers, toothbrushes, and other items needed at the rescue centers. The store also made a donation of $5,000 to SANCCOB, and staff from their Table View market served hot soups and stews to the hungry volunteers working there. Long after the crisis was over, the store continued its support of the rescue center; Peter, Pamela and Percy Pure Pilchards Pet Food was sold at Pick 'n Pay markets until 2009, with a portion of the proceeds still going to help the penguins being cared for at SANCCOB. Because of the pilchard shortage in the region, this particular pet food is no longer available, but the store hopes to launch another item to help fund the center's important rescue and rehabilitation work.

A little known fact is that, some time after the penguins arrived back on their islands, it was discovered that Pamela was in fact a male penguin, so "her" name was changed to Pamelito. He and Peter both remained elusive after their return to their respective breeding islands. Pamelito was not recaptured until September 6, when his satellite tag was finally removed. Peter kept researchers in limbo far longer. He first made landfall on the eighty-second birthday of Robben Island's most famous former resident, Nelson Mandela. While the transmissions from his satellite tag showed he had made it back to Robben Island on July 18, researchers could never find him. During the days following his triumphant return home, the stealthy penguin kept frustrated search parties on their toes. More than once, they received transmissions indicating that, just an hour earlier, Peter had been precisely where they were standing. Yet despite their best efforts to flush him out, he managed to stay completely hidden. His satellite transmitter, which eventually stopped sending signals, was never found, either. After nearly four years had passed and he had not been seen, researchers began to believe the worst. Perhaps the world's most famous penguin had died after completing his long swim home. They had to accept the possibility that he had fallen victim to a predator at sea, or perhaps he had starved or become terminally ill. They had since spotted and read the wing bands of more than 11,000 penguins that had been caught in the *Treasure* oil spill, but not one of them was Peter's. They had all but given up hope of ever seeing him again.

In the spring of 2004, Dr. Peter Barham, a professorial fellow and polymer physicist (as well as a longtime penguin enthusiast) from the University of Bristol in England, was out on Robben Island with an Earthwatch team, monitoring the African penguins. He had designed a new polymer wing band that was being field-tested on the penguins, and they were tracking the birds to see how well these new ID bands were working. The team also recorded the numbers on the metal wing bands of every penguin they saw, collecting data that would be used by local researchers to follow the breeding success and survival rates of the penguins that had lived through the *Treasure* oil spill.

On April 1, 2004, Dr. Barham was looking through his telescope when a penguin sitting on a rock in the distance caught his eye. Focusing in on the bird's wing band, his breath caught as he read the numbers: A1405 . . . Could this possibly be Peter? The first several numbers matched, but the last number on the band was obscured. Finally, after watching the penguin for fifteen agonizing minutes, the bird shifted position and the last number on its wing band came into view: A14059. It was Peter! Hardly believing his eyes, Dr. Barham asked one of the Earthwatch volunteers to take a look and tell him what identification number she saw. After she recited the same number back to him, Dr. Barham excitedly called Mario Leshoro, the environmental officer on Robben Island, and told him he was looking at Peter the Penguin. Believing it was an April Fool's Day joke, Mario burst into laughter. But Dr. Barham was insistent. Still not fully convinced, Mario hopped into his truck and drove to their location to take a look for himself. Peering through his binoculars, Mario saw that it was indeed the one and only Peter. The bird that had caused researchers years of frustration, worry, and even despair was alive! Four years earlier, the world's most famous penguin had concluded his legendary swim on Nelson Mandela's birthday. Now he had chosen April Fool's Day to make his long-awaited reappearance. Peter, apparently, was a penguin with a flair for the dramatic.

14

Thirty-five Hundred Hungry Mouths to Feed—Raising the Abandoned Chicks

The priority is definitely breeding pairs. They have a much higher survival rate. In an ideal world we'd like to save the chicks as well but, you know, we just have to deal with the resources we have at the moment and set priorities.

—ANTON WOLFAARDT, CNC AND ADU

While the adult penguins were being rescued, another drama was taking place on the islands. Because it was the height of the penguins' breeding season, thousands of chicks were still in their nests; some had just hatched, and were blind and utterly helpless, while others were almost three months old and nearing independence. On the day the *Treasure* sank to the bottom of Table Bay, approximately 6,000 chicks were being reared on Robben Island, and nearly 9,000 older chicks were close to fledging on Dassen Island. But when the adult penguins were removed from their nests to save them from the meandering oil slick, rescuers had been forced to leave the chicks and thousands of fertile eggs behind. Without help, these orphaned chicks would not survive. Sadly, there was not enough manpower to facilitate the labor-intensive process of raising all of these penguin chicks.

Baby penguins must be hand-fed every few hours in order for them to survive, and the process is even more complex and time-consuming than force-feeding an adult penguin. It's a good thing they're so adorable when they're young–it's a great strategy for getting us humans to happily do all the extra work required to raise them. Under normal circumstances the procedure goes something like this: Every morning, a nutritious formula is freshly made for the hungry chicks. This pungent concoction, with the consistency of a milk shake, is typically made up of herring, purified water, shrimp, and–based on the chick's age–a varying assortment of several vitamins. Each chick is weighed first thing in the morning, as well as before and after each time it is fed. It is then fed 10 percent of its morning body weight at each feeding throughout the day. Newborns are given formula only; prior to each feeding, some of the formula is slowly warmed over a burner or an electric plate to 96°F. A precisely measured amount of the warm purée is then drawn up into a large syringe; to avoid damaging the chick's delicate palate and tongue with the hard plastic syringe tip, a short flexible catheter is placed on the end of it. The soft catheter is put into the chick's mouth and the warm formula is given. After a few weeks of this liquid diet, small pieces of filleted fish are introduced along with the formula. Eventually, the chicks are weaned off the formula and graduate to filleted fish only, then fish halves, and finally whole fish.

Oftentimes, young chicks are reluctant to feed from their new human caretakers, so they must be properly stimulated first. To generate a feeding response, a V shape is made with two fingers, then, with the fingers pointing down, they are placed over the chick's beak and vibrated rapidly side to side, bouncing against both sides of the beak. This mimics the movements made by the chick when feeding from its parents and encourages the tiny penguin to start making swallowing motions. The chick must start gulping before attempting to feed it; if the formula is squirted into its mouth too early, the chick can accidentally inhale the thick liquid, leading to choking or aspiration pneumonia, both of which can be fatal. A small amount of formula is pushed from the syringe each time the chick responds to the feeding stimulus. As they do with their parents in the wild, chicks often rest between each gulp of food before begging and whistling for more. Once they have consumed all of the formula in the syringe, the feeding is done, even if they are still begging for more. Overfeeding can lead to life-threatening digestive problems: this is why feeding a specific percentage of their morning body weight is so important.

Clearly, the process of preparing formula every morning and feeding young chicks takes a great deal of time, skill, and patience; unfortunately, there just were not enough penguin specialists or trained volunteers available to raise so many orphaned chicks during the *Treasure* rescue effort. All was not lost, however. The chicks that had not starved to death, succumbed to the elements, or fallen victim to predators out on the islands were collected and brought to the Salt River rescue center. But, with 19,000 oiled penguins to rehabilitate, and limited resources, rescue directors and researchers had to make the very difficult decision to humanely euthanize the smallest chicks–those less than one month old. The priority would still have to be saving the adults. It was known that they would continue to reproduce after being rehabilitated, making this a scientifically sound choice. Also factoring into this decision was the fact that, in the time it took to raise one newborn chick, several adult penguins could have been saved. This harsh reality was very difficult for most of the volunteers to accept; even for those of us who fully appreciated the broader picture and conservation strategy, it was a painful verdict to live with.

Though not all of them could be saved, 2,643 abandoned chicks were eventually rescued from Robben Island, and 707 from Dassen Island. As it was well into the breeding season, many of these chicks were between one and three months old. At their larger size, they required fewer feedings per day than newborns, so a handful of people were tasked with hand-raising this group of orphaned chicks. Efforts were focused on those weighing three pounds or more, knowing that the future of the species would be best served by ensuring these older chicks survived long enough to fledge.

Of the 3,350 orphaned chicks rescued from the islands, 723 were raised by our teammates Steve Sarro and Lauren DuBois at Salt River; the rest were distributed to several wildlife rehabilitators in and around Cape Town that had prior experience raising penguin chicks.

Tina MacDonald, who had raised 800 chicks during the *Apollo Sea* rescue, cared for 1,557 of them at her wildlife sanctuary, Monty's, in Duynefontein; Cheryl Campbell, a longtime SANCCOB volunteer, reared about 500 at her home in Table View; 52 were sent to Sea Point Aquarium in Cape Town; 466 were flown to the Oceanographic Research Institute in Durban (courtesy of South African Airways); and 30 of the chicks were reared at Bayworld Oceanarium in Port Elizabeth. Due to the hard work of many dedicated individuals at these various centers, these orphaned penguins were given a second chance at life.

But someone had to make the tough call about which young chicks would get this second chance and which would not. As the boxes of chicks arrived at Salt River from the breeding islands, it was up to Steve and Lauren to make the final determination as to which chicks they believed they could raise, given the resources at hand. As they sorted through the cardboard boxes of peeping brown fuzzballs, they had to make hundreds of gut-wrenching but pragmatic decisions. Ultimately, there were 319 baby penguins whose brief lives came to an end the day they arrived at Salt River.

As painful as this was, they had to remind themselves that the number of chicks put to sleep was a small percentage (9.5 percent) of all the chicks collected for hand-rearing, and just a fraction (2 percent) of the total number of chicks born that season. Probably only 30 or 40 of the 319 chicks that were put to sleep would have survived to breeding age had they been raised by their parents in the wild, but it still was upsetting to have to put an end to these tiny lives simply because we did not have enough people to care for them. These precious victims were a grim reminder of the devastating impact of this oil spill.

In addition to the chicks that were rescued from the two breeding islands, there were about 7,000 chicks on Dassen Island that were almost fledged. Rescuers were greatly relieved to find so many chicks in this stage of development, because with their first set of waterproof feathers almost grown in they would be able to safely go to sea to hunt for themselves. Hunger would soon drive these young birds to enter the ocean to find their own food, as their parents were no longer there to feed them. While they would still have to figure out how to swim and catch fish and evade predators, their new plumage signified a critical rite of passage: once they had these water-resistant feathers, they could become independent. Though they still had bits of brown fluff here and there, they now sported the silvery-gray back feathers and off-white chest feathers of a juvenile African penguin. In South Africa, they are called Blues at this stage, because their new feathers have a bluish cast to them. (They are not the same penguins as Little Blues, which are a different species native to Australia and New Zealand.) These young Blues were at the age at which their parents would soon be kicking them out of the nest, so they were left on the island to fend for themselves, as their odds of surviving were just as good as any other fledgling's at this point. Any chicks that were still fully covered in brown fluff were brought back to Salt River.

On their second day at Salt River, Lauren and Steve helped construct the chick-rearing room and prepared for the first chicks to arrive. Although they would have normally fed formula to chicks this age back at Sea World and the Baltimore Zoo, that was not feasible here, so they decided to try raising them on solid fish. They would cut the pilchards in half and force-feed them to the chicks. They weren't sure if the young penguins' digestive systems would be able to handle solid food, but it was the only possible way they could get them all fed. The process of feeding was essentially the same as with force-feeding the adults: grab a penguin, squeeze it between your thighs, pry open its beak, and shove a fish down its throat. Next step: pray it doesn't spit out the fish. Next step: grab another fish and repeat.

Because it had been decided that the bulk of the available manpower would be devoted to saving the oiled adults, Lauren and Steve were primarily on their own in the chick-rearing room. Occasionally, they had a volunteer or two who helped them clean the pens; but, every day, the two of them caught and force-fed more than 700 penguin chicks. Before long, their hands and fingers were slashed and bloodied from the penguins' sharp beaks and toenails. Though we had all been provided with thick rubber gloves, it was almost impossible to manipulate the chicks' small beaks while wearing these bulky gloves, so Steve and Lauren didn't bother using them. But their fingers were so battered and raw from constantly prying open hundreds of beaks that they needed some protection. 3M's Vetrap became their saving grace. This stretchy bandage material, which stays on even after getting wet, is designed to stick to itself without the use of tape or fasteners, so it's ideal for use on pets and wildlife, as well as on athletes. Soon, Lauren and Steve, along with many other veteran penguin feeders, were sporting this brightly colored wrap on their injured fingers. Although hundreds of volunteers passed through the rescue center each day, we were able to quickly spot the more experienced (or perhaps more intrepid) penguin feeders by the brightly colored Vetrap they wore on their fingers–a sure sign they had cast aside the cumbersome rubber gloves as well.

Ideally, penguin chicks are fed every few hours, but this was impossible at Salt River. Upon arrival at the rescue center each morning, Lauren and Steve conducted a check of the pens in their room to be sure no chicks had died overnight, after which they began the demanding chore of feeding their young charges. With more than 700 mouths to feed, this task took from the time they walked into the rescue center until ten o'clock or later at night. Instead of being fed steadily throughout the day, as they would have been had they been raised by their parents or in a zoo or aquarium, these penguin chicks were lucky to get fed once a day. Once the chicks were trained how to free-feed, Lauren and Steve could get through their daily feeding much faster, which then enabled them to help out in other rooms at the rescue center. Also, once the chicks were free-feeding, a few experienced volunteers were allowed to come into the chick-rearing room in the evenings to offer additional fish to them after they had been fed by Lauren and Steve.

Even though they were responsible for a huge number of chicks, things were going smoothly, and the young birds were thriving under Steve and Lauren's tender loving care. After several weeks of attentive and constant fostering, the chicks had come to recognize the two of them, and the entire group would come rushing over en masse and crowd around them whenever they entered the room. Lauren and Steve frequently found themselves mobbed by fluffy brown chicks, standing at their feet with mouths agape, whistling and begging to be fed. One fateful morning, however, they entered the chick room to find that one of the pen dividers had fallen down overnight. A large number of chicks were lying on top of the collapsed section of fencing, and when they shooed the chicks off to lift the fencing and put it back into place, they made a gruesome discovery. Underneath the fallen divider were two dead chicks. They had no way of knowing how long the chicks lay trapped under there, and whether they died quickly or were slowly crushed while the other chicks clambered up to sleep on top of the collapsed fencing.

It was a disturbing scene, but by this point Lauren was so numb from the unrelenting stress and exhaustion of the rescue effort that the deaths barely registered. She wearily thought to herself, "Ach–alright. Guess we lost those two." Though she knew she should be more upset–and normally she would have been–she now felt strangely resigned and detached. This undoubtedly was a subconscious mechanism to cope with the intense daily pressure, and to manage the burden of guilt that came from holding each tiny chick in her hand and deciding its fate. If she allowed herself to feel the full weight of those painful decisions, it would have been too excruciating to show up at the rescue center every day.

Unfortunately, the volunteer who had set up the pen divider the night before was there to witness the resulting accident and was completely devastated. She began crying uncontrollably, choking out between heaving sobs, "It's my fault. I left that there. It's my fault. I'm so sorry!" Lauren tried her best to soothe and reassure the distraught young woman, telling her that she didn't hold her responsible for the freak accident. She pointed out that it could have easily happened to anyone, but the volunteer was inconsolable. Overcome with guilt and remorse, she could not stop crying and blaming herself for the tragic deaths of the two young chicks.

Of the 723 chicks under their care at Salt River, these were the only two they lost. This was a remarkable accomplishment, especially given the extreme circumstances and the untested feeding methods they were using. While the loss of the two chicks was unfortunate, their untimely deaths had to be put into perspective. If left under their parents' care in the wild, a large number of the chicks in that room would have died in the first few weeks of life—even if there had been no oil spill to disrupt the breeding season. Nature was far more cruel and unforgiving; at least, under human care, all of these chicks had a fighting chance. But thoughts of the smallest chicks they had been unable to raise were never far from their minds or their hearts.

One day, Jay Holcomb entered the chick-rearing room cradling a newborn penguin chick in his hands. It was one that Lauren had brought to the vet earlier that morning to be euthanized. Looking at her imploringly, he held the chick out and asked: "You—you can't take care of this one, huh?" already knowing in his heart how she would answer. Lauren studied the tiny chick in his palm; its eyes were barely starting to open, which meant it was just four or five days old. She knew it would be impossible to honor his request. "Well, *yeah,* Jay," she replied sadly. "I *could,* if I was able to feed him five times a day on formula, and give him a nice warm place, and have one person dedicated to him. But I can't—not if you want me to take care of all these other guys. But the question you're asking me—yes, I *could* raise this chick, but just not *here.* Not if I'm going to save these other chicks." As much as he hated to admit it, Jay knew she was right. In another place and at another time, Lauren could have easily saved that chick. She hated having to say no to him and having to decide which chicks were too small to be saved; even more, she couldn't stand knowing that she could have raised these youngest chicks if she just had enough time or help. She was deeply troubled by having to assume the role of the grim reaper, when all she wanted to do was save as many lives as she could. But it had to be done to give the species a fighting chance. Still, these agonizing decisions would haunt her for years to come.

15

SANCCOB—Thigh Biting and Other Hazards

I don't think that a week ago we expected to ever have that many penguins coming into SANCCOB oiled. I think if we all stop and think for a minute—it is very devastating, what's happened. We try at SANCCOB not to think; we're just working, trying to get the birds off [the islands]. We're looking at 41% of the entire world population of African penguins—of a vulnerable species. I think it is far more serious than a lot of people think.

—ESTELLE VAN DER MERWE, SANCCOB

It was one week into our grueling mission, and I was just starting to feel as though we had a grip on things in Room 2. We had an organized system in place for managing the care of the 4,500 penguins in our room, and a core group of regular volunteers who were now able to help us train newly arriving volunteers. Then Jay strode into our huge room one afternoon and, though I didn't realize it when I saw him approaching, my world was about to be turned upside down again. Winding his way toward me through the long rows of blue pools brimming with penguins, he finally reached where I was standing.

"Come with me," he instructed. "You need to take a break and have some food." But I wasn't hungry, and told him as much. Besides, who had time for a break? Ever since we'd first arrived at the rescue center, it truly hadn't occurred to me to eat. There wasn't time to notice hunger pains or other bodily needs—we were just too busy. Even if we had noticed our stomachs growling, there was simply too much work to do to consider stopping for any reason. (In fact, I would be 12 pounds lighter by the time we left Cape Town—nearly everyone working the rescue lost 10 to 20 pounds.) But after insisting once again that I stop to eat, he grabbed me by the arm and physically hauled me out of the room. Clearly, I was not going to have a say in the matter.

Jay wisely knew—from managing countless animal rescues over the years—that we had to eat to keep our energy up and continue at this relentless pace; without fueling our bodies, they would quit on us, and then we would be of no use to anyone. He had seen too many staff members and volunteers drive themselves into the ground during operations like this one in their fervor to save the animals. As he escorted me from our room, I stopped at the small sink located by the entryway and tried to wash the oil and grime from my hands, but as usual, it was a pointless exercise. The water was cold, there was no soap left, and the cotton hand towel was soaking wet and black with oil. Trying to wipe away some of the filth, I noticed again the thin, black line of oil embedded under each fingernail and along the rim of each cuticle. When showering each night, I tried to remove these vestiges of work, but black traces of oil and coal dust stubbornly remained until after I had returned to the States and was no longer handling oil-covered penguins all day.

As we entered the staff break room, Jay motioned to a long folding table in the middle of the room, ordered me to take some food from the serving trays lined up there, and directed me to a chair. A local caterer had taken pity on the rescue staff and sent over several large chafing dishes filled to the rim with hot, freshly cooked food. The array was impressive and a bit overwhelming. There was chicken with gravy, mixed vegetables, rice, and manicotti topped with marinara sauce. This was an unusual treat, as we never had actual meals at Salt River. On the rare occasion that we did grab something as we rushed past the Red Cross food station, it was typically a slice of bread with cheese or butter, or perhaps a cup of soup. The wonderful people who had made and donated this healthy feast wanted to be sure that, on this one day at least, we had a good, hearty meal.

Just as I had taken my first bite of manicotti, Linda entered the break room. She and Jay immediately sat down to my right and, looking quite serious, said they had something important to discuss with me. Immediately wary, I put down my fork and turned my attention to them. They told me that help was needed at the SANCCOB facility. Apparently, the penguins there were not all getting swum every day, and this was of great concern to them. Without daily swims, they weren't regaining their waterproofing, which meant their release would be delayed–and the longer these animals remained in the rehabilitation centers, the more stress it would put on them. Linda and Jay also worried that the people at SANCCOB might not be getting all of the resources they needed, but they didn't elaborate.

"We would like to send you over there to help supervise the rehabilitation and swimming of the penguins," Jay said. "We have been observing your work in Room 2, and believe you have the strength and skill to be successful at SANCCOB."

I was flattered–and also a bit taken aback by their faith in me. I also wasn't completely clear about everything they expected me to do at this other site. I knew that swimming all of the penguins every day was a priority, but they hadn't gone into detail about anything else, and the rest of it seemed rather ambiguous. All of which made me rather uncomfortable.

"You don't *have* to go," Linda said. "We know that you've just settled into a routine here, and we aren't going to force you to make this transition. It's up to you. But we really think you could help them out, so we hope you will choose to go. If we weren't confident that you could make a difference there, we wouldn't be asking you do this."

"When do you need a decision?" I asked.

"Well, *now* would be good," answered Jay. Then, grinning slyly, he added, "No pressure, though."

I had been in South Africa for just one week, and now they were asking me to take on a new, vaguely defined responsibility that I wasn't certain I was prepared for. Feeling quite anxious (and far less sure of myself than they seemed to be) but willing to do whatever was needed to help the penguins, I agreed to make the move to SANCCOB. When I asked when the transition would take place, they said it would be the next morning. I felt my stomach drop. Instantly I felt a great deal of pressure; the end goal was critical, their expectations were high, and I desperately did not want to let anyone down, least of all the penguins. I was also hesitant because this transfer meant that I would no longer be working alongside my teammates, to whom I had grown close in our short time together. By the time we had finished our conversation, the food on my plate was cold. It really didn't matter, because I no longer felt like eating.

The next morning, Jay and Linda sat with me during breakfast so they could impart some last-minute advice and instructions. In addition to my helping to get the birds swum every day, they wanted me to observe the procedures at SANCCOB and provide recommendations regarding animal care or staffing that I believed would help them meet their rehabilitation objectives, as well as suggest any additional resources I thought they might need. As I would be walking into an unfamiliar facility with no knowledge of the setup, this seemed like a tall order, and I realized I would have to figure things out very quickly. Though still apprehensive, I promised Jay and Linda that I would do my best.

After bidding my teammates goodbye, I climbed into the car that would take my new co-workers and me to SANCCOB, located about a mile away, in Table View. I was deeply saddened to have to part company with my team at this point. I knew I would miss each of them, as well as the support and camaraderie our bond provided. We pulled away from the Dolphin Beach Hotel, and, after a short ride, turned onto a side street off the main road. Soon, we saw long rows of parked cars lining both sides of the street. At a bend in the road, our driver pulled into a parking lot filled with trucks, trailers, and camper vans covered with tarps. Dozens of people were rushing about, and it looked like the base camp for some sort of military operation.

So this was it. I had finally arrived at the world-renowned SANCCOB. Because of my work with penguins at the New England Aquarium, I had heard a great deal about this organization over the last few years, and I knew they were the premier penguin rescue center in South Africa. But the facility was much smaller and far less impressive than I had imagined. A simple wooden sign, which looked like it had been built in someone's basement workshop, announced the entrance to the center. It consisted of two pieces of plywood cut into cloud shapes and layered one in front of the other. The sign was painted white, adorned simply with a drawing of two African penguins in the upper-right corner and a black line near the bottom, under which "sanccob" was written in rounded, lowercase letters. A low-slung olive green building sat to the right of the sign, and a cement walkway directly below it drew us into the space. Gazing to the left, beyond the parking lot, I could see our hotel in the distance across the sprawling marsh that bridged the expanse between the two locations. With the morning mist still rising from the moist grassy earth, the vast structure on the horizon appeared to be veiled behind an enormous shroud of gray chiffon.

As soon as we entered the rescue center, I was introduced–and quickly handed off–to Sam Petersen, whom I was told was the bird rehabilitation manager. Although it was immediately clear that she was stressed out and being pulled in many directions at once, she was nonetheless warm and gracious as she welcomed me to the center. She was very attractive, in her early twenties, with short, strawberry-blond hair and skin lightly dappled with freckles. What stood out most, though, was her husky voice. She sounded like a sports coach who'd gone hoarse from yelling at her team for days on end. I'll never know if it sounded that way from sheer exhaustion and the constant strain of giving orders to workers during the rescue effort, or if her voice always had that same raspy quality, but it was quite distinctive. After introducing me to some of the staff members and volunteers I would be working with, Sam gave me a tour of the grounds so I could orient myself.

The main hospital building was to the right of the entrance, while the open-air holding pens and swimming pools covered about one-half acre on the left. Cement walkways connected the pools and holding pens, which were delineated by aluminum chicken-wire fencing and gates. The holding pens themselves had cement floors, which were not ideal for the penguins' tender feet, but the hard flat surfaces were easy to hose down and clean. During this rescue, in particular, with approximately 400 penguins crammed into each pen, being able to readily wash away the copious amounts of guano they produced was essential. I noticed that one of the pens near the back had a small group of birds separated from the rest. Sam told me these birds had been pulled aside because they had bleeding ulcers. When I asked how they had arrived at this diagnosis, she pointed to their guano, which, instead of being white and green, was a solid chocolate brown. The change in color was from digested blood, as well as ingested oil. When birds were seen excreting this dark brown guano, they were separated out and brought to the ICU for further treatment. In addition to the nine holding pens, there were four shallow pools, one of which was set aside exclusively for several birds that were permanent residents there. This included a collection of about two dozen penguins, cormorants, gannets, pelicans, and gulls whose injuries had proven too severe for them to survive on their own again in the wild, so they lived at SANCCOB and served as educational birds for the local schoolchildren who visited the facility.

Inside, the small building was separated into several areas. There was a large room on the right end that normally served as an educational classroom for visiting students—however, it was now being used as a hospital area for ill or injured penguins. There was a second, smaller ICU on the opposite end of the building, where more critically ill birds were treated. There was also a small office area, two cramped toilet stalls, a laboratory for conducting blood work, and a large open room in the middle where the penguins were washed. Behind the building, in a partially enclosed hallway, were freezers for fish storage, as well as a food preparation area where fish was thawed and vitaminized for the penguins. Beyond the far end of the main building, a temporary open-front shelter had been erected to house a food and drink station that was staffed by local volunteers. Several pleasant women, who looked to be in their sixties and seventies, were there that morning, handing out refreshments and offering encouragement to the weary workers.

After completing our tour of the facility, the lead rescue team members gathered in Estelle's office for a brief meeting to review the staff organizational chart and discuss the rehabilitation strategy. Overseeing the entire rescue effort was Estelle van der Merwe. Although she normally served as center manager of SANCCOB, she had temporarily stepped out of that role in order to function as Treasure Oil Spill crisis manager. As such, she was the point person for nearly every aspect of the rescue–from bird care to logistics, to conducting media appearances, and overseeing strategic planning meetings with conservation scientists and bird care experts. As the lead person during this crisis, Estelle essentially carried the full weight of the rescue operation on her shoulders. It was an awesome responsibility.

Taking over Estelle's usual role of centre manager for the duration of the rescue effort was Karen Trendler. Karen had her own wildlife rehabilitation center in Pretoria, where she rescued and rehabilitated many species of animals, including hippos, rhinos, elephants, and warthogs. As the founder and director of Wildcare Africa Trust, she had been devoted to the care and protection of African wildlife for sixteen years. During that time, she had also worked as the crisis manager for several oil spill responses, and had received national awards acknowledging her important work saving injured and orphaned animals. Karen's primary duties during the *Treasure* oil spill crisis were to oversee operations at SANCCOB, and to order most of the necessary supplies and equipment for the two rehabilitation centers, which included everything from medicines to blood-testing supplies to fencing to pilchards for the penguins. Next in line was Sam Petersen, who–like me–was a veterinary nurse, and SANCCOB's bird rehabilitation manager. I was then told that I would be working under Sam, with the title of "bird rehabilitation supervisor."

Also present at this meeting was Phil Whittington, a PhD student with the ADU who was conducting research for his dissertation on the demography of African penguins. He was keeping track of all the penguins rescued and rehabilitated during the recovery operation. On a daily basis, he recorded how many oiled birds were brought to each rescue center, how many were washed, how many died, how many were euthanized, and how many were released. The data he compiled would not only provide a statistical record of the direct impact of the *Treasure* oil spill; it would also prove helpful to scientists studying the long-term effects of oiling and of the rehabilitation process on African penguins.

After this brief meeting, with little other instruction, I was sent outside to oversee a swimming session for a pen of penguins. The birds were swum in three tiled pools, each filled with about two feet of fresh water. The pools were fairly small (about 20 square feet), so with 400 penguins in each pen, the birds had to be swum in groups of about 100 at a time. In an attempt to get more penguins swum, I decided to include more than the usual number of birds during my first swimming session. To swim a group of penguins, several volunteers would enter their holding pen and using lengths of chicken-wire fencing about six feet long and three feet high, they would herd a portion of the birds out of the enclosure, down the cement walkway, and into the waiting pool. Anxious to get away from their human handlers and into the water, the penguins typically began shoving each other out of the way as they approached the pool.

Plunging en masse into the shallow pools, they would immediately start to swim and clean, flip-flopping about on the surface of the water like exuberant puppies rolling around in the grass. Barrel-rolling and brushing their wings rapidly back and forth across their sides to bathe themselves, their stubby tails wagged vigorously from side to side as they zipped and zagged in erratic patterns through the water. Though they attempted to dive, the pools were too shallow to do so effectively. Penguins are more buoyant that close to the surface, which makes it difficult for them to stay submerged, so the birds could only glide just below the surface for brief spurts. These daily swims were the only time they exhibited typical penguin behavior or appeared to be enjoying themselves at all. Although it was rewarding to see them in the water, there was a slightly frenetic quality to their swimming–it seemed as though they believed that, by swimming fast enough and erratically enough, they could avoid us and burst through the walls of the pool to escape back into the open ocean nearby.

With more than 3,000 penguins to swim and only three pools, it was impossible to swim all of the penguins every day. It took time to herd the birds in and out, and the pools had to be drained and scrubbed clean after each session before being refilled for the next group of penguins. These daily swims were essential for several reasons. First, the guano (their own and from the other penguins) was removed from their feathers as they swam; if the highly acidic guano remained on them, it would cause feather rot, which would prevent them from being able to waterproof their feathers until they molted again. As penguins only molt once a year, it was vital for their feathers to stay in good condition. Second, the penguins could drink while in the pool, which helped keep them hydrated. Third, the exercise was beneficial for reducing stress, and for keeping their swimming muscles in good condition. Finally, and most critically, their daily swims helped to rinse any remaining soap residue from their feathers. This was essential, because any residual soap would interfere with their ability to make their feathers watertight again. To ensure the penguins stayed in the water long enough to exercise and cleanse their feathers, workers stationed by the pool exit ramps waved and shook yellow oilskin jackets whenever the penguins approached; this startled the penguins, making them swim away. The birds had to leave the pool eventually, however, so it could be readied for the next group of penguins.

To shepherd them out of the pool, one or two people would wade into the water to try to corral the birds and get them to move toward the exit ramps. At the top end of each ramp was a gate with a low wooden door that swung open to let the birds in and out. Usually, the penguins started to panic at the sight of a human in their midst, and hurtled themselves away from us. Most made a direct beeline for the exit ramp, but there were always some that would start zipping around chaotically in the pool. It was while trying to get the large group of penguins out of a pool my first morning at SANCCOB that everything quickly unraveled.

After walking through the open gate, I stepped down into pool #1 and sloshed my way toward the swirling mass of swimming penguins. Disturbed by my presence, the birds quickly shot away from me in all directions, so I moved to the opposite side of the pool, from where I could begin herding them toward the exit ramp. Their stiff wings slapped and beat against the surface of the water as they tried to evade me. Most of the penguins bolted for the open exit gate, where they scrambled out. About a dozen frenzied penguins, though, were bunched up in the corner by the exit ramp, flailing about and clambering on top of one another in a desperate attempt to get out of the pool. Several were wedged behind the swinging door that opened in toward the pool, and they were splashing, honking, and biting each other in their panicked state. As they frantically climbed on top of each other, some penguins got pushed underwater and became trapped beneath other birds. Fighting their way back to the surface, they came up sputtering and gasping for air. Their breath came in loud bursts and deep, staccato coughs as they choked on the water.

Terrified that they were going to start drowning, I hurried over as quickly as I could through the knee-deep water and plunged my hands into the fray, trying to grab a few penguins, so I could pull them out. I could hardly believe that they were on the verge of drowning when they were just inches from dry land, but they seemed too panicked to realize that the exit ramp was right there next to them. I was horrified as I thought about penguins potentially dying or coming down with aspiration pneumonia. If any of them died as a result of this swimming session, I would never be able to forgive myself. These poor birds had been rescued, cleaned, and rehabilitated, and were now just minutes from drowning in two feet of water–and all while under my watch!

It finally dawned on me that the volunteers standing just beyond the exit ramp were frightening the penguins, causing them to remain in the water where they felt more secure. Their presence was also disturbing the penguins that had already left the pool; those birds were now huddled in the open gate, blocking the egress of the birds still in the water. I told the volunteers they had to move away from the gate so the penguins would feel safe enough to move forward through it, allowing the other birds behind them to get out of the water. By now, several of the penguins in the pool were in a state of full-blown panic, and I was losing my composure as well. When the volunteers didn't back away fast enough, I started yelling at them to get out of the way. There were already several penguins in SANCCOB's ICU being treated for aspiration pneumonia, and the last thing I wanted was for some of these birds to be added to that group of gravely ill patients.

Finally, the people beyond the exit ramp moved away from the open gate, the group of penguins gathered there started walking away from the pool, and I was able to get the rest of the frenzied birds out of the water. The chastened volunteers herded the penguins back into their holding pen while I climbed out of the pool, incredibly relieved that the disastrous swimming session was over. But I was mortified by what had just transpired. I was supposed to be the expert, and I had just made a complete mess of my first swimming session. I'd come close to having a few birds drown–it was still possible some of them would later become sick from inhaling the water–and it took me longer than it should have to realize why some of the penguins were reluctant to head up the ramp at the end of their swim.

I wasn't proud of how I'd handled the situation when I saw the penguins were in trouble, and felt badly about losing my cool with the volunteers. Although humiliated and defeated, I had to get over those feelings quickly in order to be effective and learn from the mistakes I'd just made. I realized that I had tried to swim too large a group of birds; that approach had failed miserably. Although the main purpose of my being there was to get more penguins swum, clearly I would have to come up with a different strategy to prevent another embarrassing debacle while attempting to reach that goal. I would also have to remember to look at things from a penguin's point of view!

My day did not improve much from there, and I felt like I couldn't do anything right. I spent the remainder of the day swimming pens of penguins and helping out wherever else I could, but I struggled to absorb and adapt to the procedures at this facility. The routines were vastly different from those at Salt River; the people were different; and my duties were different. My uncertainty and frustration mounted with each passing hour, and my mind spun as I mentally groped for something familiar, something of which I was sure. Despite my growing anxiety, I tried to remain hopeful and helpful. The people were nice enough. But I felt lost and miserable and out of place at SANCCOB, and by the end of the day, I was praying that I wouldn't have to return. Even though the daily routine there was more grueling and exhausting than at SANCCOB, at least at Salt River I knew exactly what had to be done and how to accomplish it. And at Salt River, I felt I was actually doing some *good.* At SANCCOB, I was not at all convinced that I would be able to have a positive impact.

Later that night, in the bathroom for my ritual cleansing before climbing into bed, I stripped off my damp, soiled clothes and caught a glimpse of my naked body in the mirrors covering the walls: I was startled to see dozens of raw, red slashes and multicolored bruises in various stages of healing. From chest to ankle, I bore the markings of dozens of penguin beaks. Staring at my battered body in disbelief, it occurred to me to take a photo. I thought I should document this dramatic physical evidence of our efforts here in South Africa. The wounds and battle scars marking my body were hard-won, and I did not mind getting them. In fact, I had become quite accustomed to being bitten, wing-slapped, and scratched by uncooperative and frightened penguins. One had to enter a sort of Zen state of mind about it; it would have been impossible to do our work if we stopped each time we were attacked to examine our injuries. Already I had learned to ignore each bite as it was inflicted, and just keep forging ahead with the work at hand, putting the sharp, stinging pain out of my mind.

I went back into the bedroom, grabbed my camera from my backpack, and returned to the bathroom to capture the reflected image of my body in the mirrors. But as I raised the camera to my eye, I was suddenly stopped by a sobering thought: What if, when the pictures were developed, someone at the photo lab saw this photograph and notified the police? (This was before I had a digital camera, so I was shooting with film.) After all, it looked as if I had been attacked–and I had heard of several cases where a husband or boyfriend had been arrested for assault and battery after photo labs brought incriminating photographs to the attention of the authorities. Not wanting to put my partner, Marc, at risk, I decided not to take a picture. In my exhausted state of mind, it did not occur to me that the rest of my photographs, showing thousands upon thousands of oiled penguins, would absolve him of any wrongdoing.

It is something I now regret–not taking that one photograph. Though my arms and hands are permanently lined with dozens of small white scars, I still wish I had just one photo of what my body looked like during those few weeks. Even now, thinking back on this incredible event, it sometimes seems like a dream or something I saw in a movie. The whole experience was so surreal that, at times, I feel as though it didn't really happen to me. For this reason especially, when the memories and the immediacy of the event start slipping away, that one picture would have been tangible proof that I actually *was* there, that I *did* participate in this historic event, and that thousands of penguins really *did* inflict that much damage to my body.

The next morning, after another night that passed impossibly quickly, I joined the other members of my team in the restaurant for breakfast. They would soon be heading to Salt River, and I desperately wished I could go with them. But it was not to be. Before I could finish my meal, Jay, Karen, and Linda beckoned me over to their table and asked me what resources I thought were needed at SANCCOB and what changes, if any, I thought should be implemented there. With just one day under my belt—and not a good one at that—I shared my observations with them, hoping they would be of some value. From what I had seen, it was clear that the most pressing need was for more pools. There simply weren't enough to rotate through every one of the 3,000 birds at that location each day. More short sections of fencing were needed as well to corral off groups of birds within each pen, and to help herd the birds to and from the swimming pools. While this was what I believed would make the greatest difference, I noted a few other supplies and changes I thought would help before heading outside to catch my ride to the rescue center.

Having had a rough first day at SANCCOB, I tried to recalibrate mentally. I knew I would have to adjust my attitude and my approach if I was going to be staying there. I was still feeling adrift and miserable as I struggled to adapt to the new location and, unfortunately, my frustration occasionally surfaced in inappropriate ways. Knowing I would be standing knee-deep in water for most of the day, and hoping to stay dryer than I had the day before, I had worn light cotton shorts under my oilskins instead of my usual jeans. At the end of one of the swimming sessions, I was wading into the water to shoo the birds out of the pool when one of the penguins, agitated by my presence, suddenly latched onto my inner thigh about eight inches above my knee with its razor-sharp beak and refused to let go. Pain and adrenaline instantly shot through my body. Without the extra protection of thick jeans, this crushing bite in such a sensitive spot was excruciating. The intense pain, coupled with my frustration, exhaustion, and angst, caused me to lose all control and I exploded, letting loose a string of rather crude swear words, shocking my new co-workers.

In its frenzied state—and alarmed, no doubt, by my loud outburst—the penguin started wing-slapping me for all it was worth, splashing water everywhere. Soon, I was completely drenched. I tried in vain to pry the penguin's beak open, but the enraged bird remained stubbornly attached to my left thigh. This, of course, made me curse even louder and more profusely. While I'm not shy about swearing when I'm with friends, I'm always professional and appropriate while at work or in public settings. But this time my aggravation and pain got the better of me. Everyone I met in Cape Town was exceedingly polite and even-tempered, and I'm quite sure I offended many of the people at SANCCOB with my vulgar outburst. After this incident, some of the volunteers seemed even more wary of me.

I made a deliberate effort that afternoon to smooth over anything I had said or done to offend anyone, and did my best to refrain from any further swearing. After all, I did not want to insult any of these people who were working so hard to help the penguins. By the end of that second day, there were small indications that the tide was turning. I was finally becoming more familiar with the procedures at SANCCOB and people were even starting to come to me with questions.

When I arrived the third morning, I was amazed to find that four new swimming runs had been built overnight; they were adjacent to the facility in the left-rear corner. Mariette and her team had worked straight through the night to get these additional pools built to help us get all of the penguins swum every day. Even though they were smaller than the three permanent pools, and fewer birds could be swum in them at one time, having these extra runs gave us the ability to swim all of the penguins at least once per day–and some of them twice. With these new pools installed, and a strict new swimming schedule in place, we were able to swim the penguins more frequently, which meant they would regain their waterproofing and be released more quickly.

Fortunately, things clicked for me on this third day. I had finally learned the routines, settled in, and regained my former confidence. And, much to my surprise and delight, both staff and volunteers began to regularly approach me to ask for advice and direction. Now that I had a clear understanding of what had to be done and how to do it, and had finally gained the staff's trust, I felt comfortable proposing a few changes regarding some of the husbandry proce-

The damaged iron-ore carrier MV *Treasure,* prior to sinking in Table Bay on June 23, 2000. (Mark Hutchings, *Cape Times*)

A group of heavily oiled penguins at SANCCOB, shortly after being rescued from Robben Island. (Tony van Dalsen, Department of Agriculture, Forestry and Fisheries, South Africa [DAFF, formerly MCM])

Workers herding and capturing penguins on Dassen Island. (Tony van Dalsen, DAFF)

Penguins being loaded onto an army helicopter for transport to rescue centers on the mainland. (Tony van Dalsen, DAFF)

Some of the 19,506 clean penguins after their rescue from Robben and Dassen Islands. (Tony van Dalsen, DAFF)

Loading a sheep-shipping truck with clean penguins to be released at Cape Recife in Port Elizabeth, 560 miles from Cape Town. (Tony van Dalsen, DAFF)

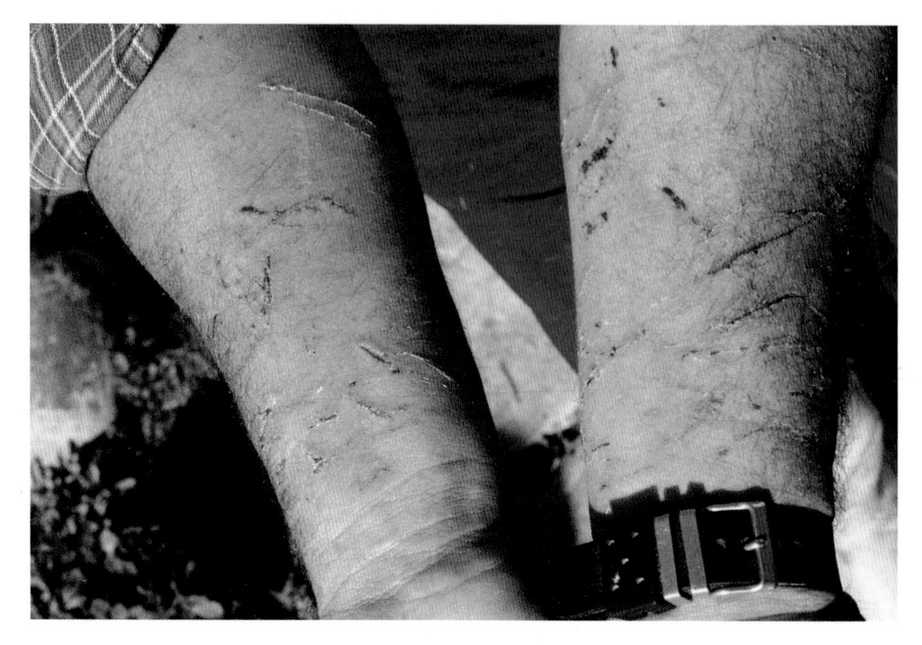

War wounds from catching penguins. Everyone was covered head to toe with bites from their razor-sharp beaks. (Tony van Dalsen, DAFF)

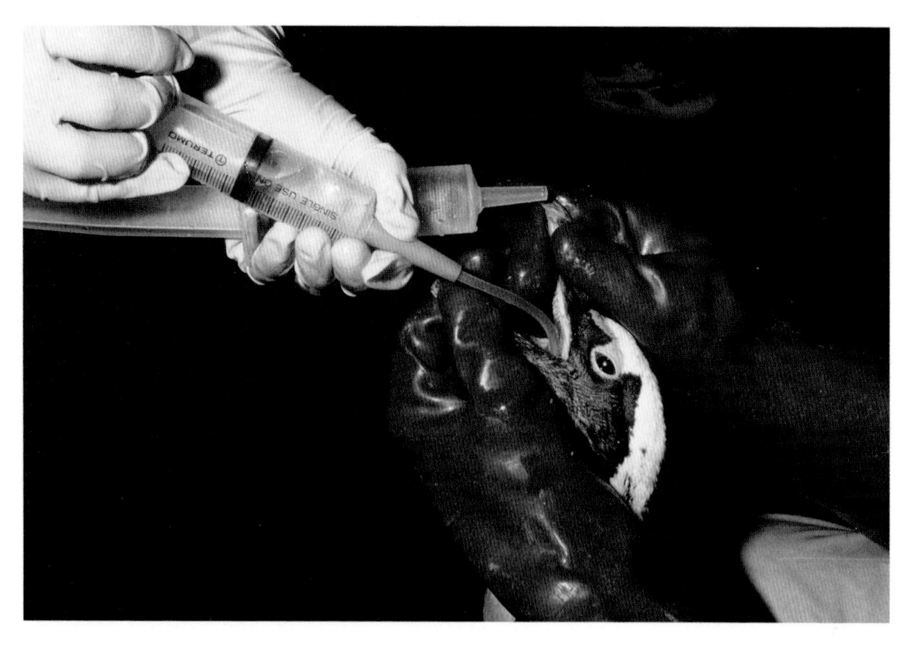

Every penguin was tube-fed an electrolyte solution (called Darrows) to rehydrate it. (Tony van Dalsen, DAFF)

Our room (Room 2) at the Salt River Penguin Crisis Centre. By this point in the rescue effort, many of the penguins had been moved outside. (Dyan deNapoli)

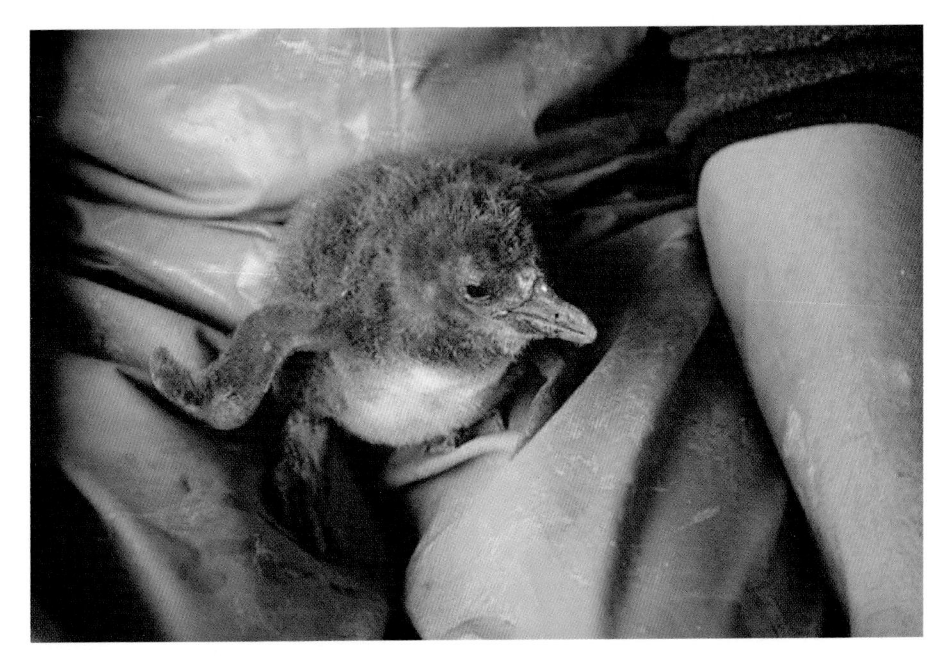

One of the young African penguin chicks. (Photo courtesy Dyer Island Conservation Trust)

Preparing some of the five to ten tons of fish the penguins consumed every day. (Martin Vince, Riverbanks Zoo & Garden)

A few of our most dedicated volunteers in Room 2 force-feeding the penguins (*left to right in pool*: Erwin, Isabella, and Dennis). (Dyan deNapoli)

A heavily oiled penguin–note the red, irritated skin around the eyes and the clumped, separated feathers. (Les Underhill, ADU, UCT)

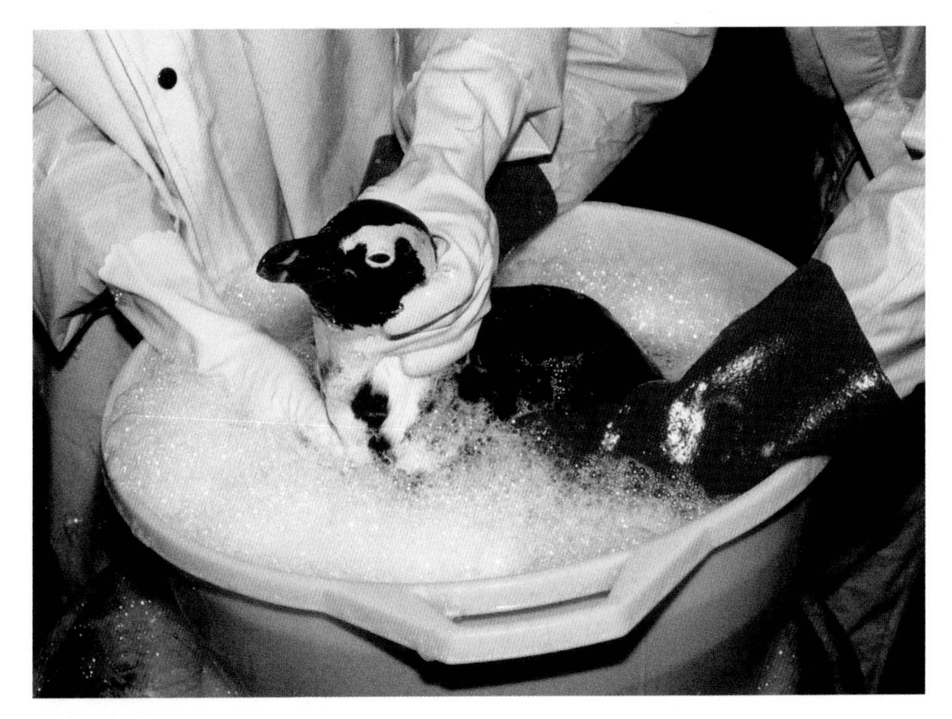

Washing a penguin with hot water and LDC dishwashing detergent. On average, it took 45 minutes for two people to clean one penguin. (Tony van Dalsen, DAFF)

Hand-raised penguin chicks upon release on Robben Island. The confused young birds stood staring at their human caretakers, not sure of where to go or what to do. (Tony van Dalsen, DAFF)

The release of a group of cleaned and rehabilitated adults at Milnerton Beach in Cape Town. The pink spots sprayed onto their chests–which fade over time–help researchers spot the penguins after they return to their islands. (Tony van Dalsen, DAFF)

dures. I was able to make some suggestions that were implemented, and things began going more smoothly for me. By this point in my stay, my working relationship with the people at SANCCOB had strengthened. The staff members, volunteers, and I were finally getting to know each other better and, just as at Salt River, feelings of camaraderie and mutual respect were developing between us.

It was a tremendous relief to have found my footing, and at the end of my nine-day stint, when they begged me to stay longer, I was shocked and pleased. I was very happy that, after my shaky start, I was able to finish my time at SANCCOB on a positive note. In the end, I was also grateful that we had found a way to work together and keep our focus on the most important goal of all: to save the penguins.

As I became an accepted member of the team, people began sharing more "insider" information with me. One afternoon several days into my stay, one of the workers pulled me aside and told me that she had a secret to tell me. I sidled up to her, the whirring clothes dryer squeaking behind us, and leaned in, curious to hear what tidbit she had to share. Apparently, she had been told that one of the penguin researchers who had been monitoring the penguins for several years on one of the breeding islands had made off with a few small penguin chicks. Although much younger than the other chicks that were being taken in for hand-rearing, the researcher apparently had a special place in his heart for the parents of these chicks, and couldn't bear the thought of their newborn babies dying alone in their nest or being euthanized.

So, unbeknownst to the other rescue workers, he allegedly smuggled these few chicks off the island and brought them to one of the offsite chick-rearing facilities to be raised by human caretakers. I like to think that they survived to fledging and were returned to their breeding islands—perhaps they have even raised several of their own offspring in the time since the oil spill. Years later, when I asked the researcher whom I suspected might have been the one who saved these chicks, he knew nothing about it. Whoever the secret rescuer is, I imagine he shares my hopeful fantasy about the chicks' fate.

As we had seen at Salt River, there were always volunteers and other folks who hoped to help us by coming up with novel approaches for carrying out specific tasks or solving particular problems. The Australian volunteer in Jill's room who wrestled a sweater onto a penguin was one such example. One afternoon at SANCCOB, Sam had a similar encounter with an older woman who lived near the rescue center. Knowing that the penguins inflicted nasty bites, this woman had invented a rather unusual penguin-catching device designed to protect our hands and arms. It was a broomstick, to which she had attached an orange traffic cone, so that it looked like a giant candlesnuffer. The woman eagerly presented the contraption to Sam, informing her that she could make more of them if needed. After looking over the curious device, Sam thanked the woman and gave her the standard brush-off line: "We don't use untested devices or methods in the midst of a crisis, but we would be happy to assess it after the oil spill rescue for possible future use." While it was clever in concept–and just the right size to trap an African penguin–it really wasn't practical, because to control a penguin and not be bitten by it, you have to grab it by the back of its head, which would have been impossible with a traffic cone over its entire body. We had a few good laughs though, as I took several photos of Sam in action while she pretended to capture penguins with her new-fangled penguin catcher.

And just as we had experienced at Salt River, the outpouring of concern and help from people from around the world was absolutely breathtaking. I met individuals from many different countries and all walks of life at Salt River, and the same was true at SANCCOB. There was no shortage of compassion and caring at either center, and there were certain volunteers who stood out, and whose dedication and work ethic made me weep with gratitude. Dianne De Villiers and Gary Egrie were two such people whom I met at SANCCOB. Dianne was a teacher and photographer, and her fiancé, Gary, had just been accepted into veterinary school. Dianne, a stunning redhead with a broad smile (I later learned she was a former high fashion model), was originally from South Africa, and she and Gary lived in the Maryland area. The two had flown in for a vacation before Gary began his studies. But, as fate would have it, they arrived the very day the *Treasure* went down in Table Bay.

On their first afternoon in Cape Town, they rode the cable car to the top of Table Mountain to enjoy the expansive view of the city and the bay. From their vantage point high on the mountain peak, they could clearly see Robben Island a few miles away; but they were horrified when they noticed a huge oil slick floating on the surface of the ocean near the island. Watching the news that night, they learned that oiled penguins were showing up on the shores of Robben Island, and that SANCCOB had put out a call for volunteers. Dianne and Gary immediately knew what they had to do. Having grown up near Cape Town, Dianne was familiar with SANCCOB, so she and Gary showed up first thing the next morning to help.

They had planned this trip home to visit Dianne's family, whom she hadn't seen in six years, and so that her parents could finally meet Gary. But as it turned out, they never spent any time with her family. Instead, they spent sixteen to eighteen hours a day in the holding pens at SANCCOB surrounded by thousands of oiled penguins. At the end of each exhausting day, they climbed into their car and drove for two hours to get to Dianne's parents' house. After arriving around 1 a.m., they showered, slept for three or four hours, then got up and drove the two hours back to Cape Town again. They carried out this grueling schedule for six solid weeks. Their original trip to South Africa was supposed to be for two weeks, but they ended up going to the airport three times to change their tickets and extend their stay. (Aware of the penguin crisis in Cape Town, the airlines waived the usual fees for changing their tickets. In fact, South African Airways provided plane tickets for our team, as well as for the other AZA penguin caretakers who flew to South Africa after us, at half price.)

Once Dianne and Gary began volunteering at SANCCOB, they found it impossible to leave; the need was just too great and they felt compelled to stay as long as they possibly could to help. They pushed themselves so hard that, by the end of their stay, they had each lost more than twenty pounds. Dianne even ended up with intestinal parasites, which she most likely contracted from an infected penguin. The two of them only left South Africa because Gary's classes were about to start.

The *Treasure* oil spill wasn't Dianne's first time caring for oiled penguins. As a young girl, she had been on vacation with her family at the shore when they came across a penguin that was covered with oil. Her father, an avid bird lover, couldn't just leave the penguin on the beach to die, so he wrapped it up and brought it to where they were staying. He had a lot of experience rehabilitating birds, so he put the penguin into the bathtub and, using dishwashing detergent, washed the thick oil from its feathers. Dianne and her sister went to the beach every day to catch fish to feed to their patient, and together the family nursed it back to health. At the end of their vacation, they tried to release the young bird, but it had apparently become attached to them during the time they cared for it. They repeatedly set the penguin by the water's edge, only to have it enter the ocean and start to swim away, then turn around and head straight back to shore. Exiting the water, it would race right over to Dianne and her family. Despite their persistent efforts, it refused to take to the sea. Concerned that the penguin would not fare well on its own, her father brought it to SANCCOB. The staff there assured him that their penguin would be released with a group of other rehabilitated penguins; in the company of the other birds, they believed that it would do just fine.

Under her father's loving tutelage, Dianne's wildlife rescue experience had begun at the tender age of ten with a penguin in their bathtub. Now, twenty-five years later, after volunteering for just a few days during the *Treasure* rescue effort, she had been entrusted with overseeing the care of 400 oiled penguins that were crammed into one of the holding pens at SANCCOB. Her father even joined her for a few days, hosing down the oiled birds throughout the day so they would not overheat in the hot African sun. Neither Dianne nor her father could have imagined that the experience they shared rehabilitating that one penguin during her childhood would culminate all these years later in their participation in the world's largest rescue of oiled penguins.

Toward the end of my stay at SANCCOB, I was assigned to spend the day with Gary, taking blood from all of the penguins in pen #9 in order to determine their fitness for release. With 400 penguins to bleed, I mentally prepared myself for a long day. The blood samples would be analyzed at the lab on site, to test the birds for dehydration and anemia, as well as for blood parasites. To run these tests, we needed just enough blood from each penguin to fill a capillary (microhematocrit) tube–a tiny glass tube about three inches long and as thick as a strand of dry spaghetti. To extract the small amount of blood from each bird, a needle was inserted into the metatarsal vein in the penguin's foot. The problem was that this minute vein is invisible to the eye or touch. It is located between the toes on their webbed feet and is found solely by knowing where it lies. So this process is what we in the industry call "a blind stick." And nobody likes a blind stick–least of all the animal that is on the receiving end of it, because it generally entails some probing and poking around with the needle before finally hitting the hidden vein.

When I worked as a veterinary nurse at a small animal hospital years before, my co-workers nicknamed me "Vampira" because I had a knack for getting blood from the tiniest of veins on the tiniest of kittens and from the threadlike veins running through a rabbit's ear. There was a definite thrill in hitting an invisible vein and a sense of satisfaction watching blood fill the hub of a needle once it had hit its intended target. It was a challenge I never grew tired of. At the New England Aquarium, staff from the Veterinary Department drew blood from the penguins while penguin staff restrained the birds, so I never had the opportunity to draw blood on a penguin; but I had always wanted to try. Initially, I was a bit unsure about finding this tiny vein in the penguins' feet. This was not a blood-drawing technique used at the aquarium. Because we generally needed more blood for our tests, we had to get it from a much larger vein, so we used the jugular. But for the amount of blood needed for the testing at SANCCOB, the small vessel in the penguins' feet would do.

Gary demonstrated the technique they were using–it truly was a blind stick–and I made several fumbling attempts before finally hitting my first elusive vein. Because such a small amount of blood was required, we just used a needle without the syringe attached. Once placed in the vein, the blood slowly filled the hub of the needle–although we sometimes had to squeeze the penguins' toes to get the blood to flow. Once the hub was full, we withdrew the needle and inserted one end of the capillary tube into the blood in the needle hub. Capillary action then sucked the blood up into the thin tube. It took about a half hour and several penguins before I became adept at finding just the right spot on the foot, but soon I was hitting the vein with regularity, and the blood-drawing process went more quickly. To keep track of each individual, every penguin had a temporary ID band placed on its wing when it was admitted to SANCCOB–and each white plastic tag had a unique number written on it in indelible marker. A "recorder" was stationed in the holding pen with us. This person recorded the ID number of the penguin we had just drawn blood from on a piece of paper on his or her clipboard, then taped the capillary tube next to its number; each sheet was then brought to the lab for testing once it was full.

To restrain the birds for blood drawing, the handler first had to catch a penguin, then position it with its back pressed against his own stomach, the bird's chest facing out toward the person drawing blood. Some penguins immediately froze when picked up, an instinctive reaction exhibited by some wild animals when handled by humans, but most kicked and struggled, trying desperately to escape being held. There was one special man, though–a volunteer named Welcome–who truly had a magical touch with the penguins. Welcome was gentle, kind, and caring; like many of the volunteers at SANCCOB, he was a member of the Xhosa tribe. The Xhosa people traditionally raise cattle, and Welcome had inherited his father's farm when he was five years old; just over thirty at the time of the oil spill, he'd had a lot of previous experience working with animals. His serene, Buddha-like presence had a wonderfully calming effect on the birds: the moment he picked up a penguin and placed it against his stomach, it instantly relaxed. (I swore I could hear the birds letting out a sigh of relief when he lifted them and cradled them gently in his long, slender arms.)

Everyone else handling the wild penguins had to grab on and hold them very tightly, which is generally what is required to avoid being bitten. But Welcome had the magical ability to hold the penguins with the lightest of touches–using practically no restraint at all–and the birds just sat there calmly, without moving, and allowed us to poke needles into them without complaint. It was the most astounding thing I've seen. Even when first pricked by the needle, the birds never jumped or flinched as most penguins did. I quickly dubbed this remarkable man "Welcome, the Penguin Whisperer." I was in awe of his ability to soothe these wild birds, and I begged him to be my handler for any blood drawing I had to do during the rest of my time at SANCCOB. Working with Welcome was such a pleasure and an honor; not only did the penguins respond positively to him, but his sunny, warm disposition made him a joy for us to be around as well. His eyes sparkled with kindness and compassion, and a large, easy smile spread across his handsome face, lighting it from within. Soft-spoken and humble, his presence was a gift to the penguins and to all who had the privilege of working with him.

Years later, in the spring of 2005, one of my former colleagues from the New England Aquarium visited SANCCOB and returned with a short video clip he had taken of Welcome, who had since become a valued staff member there. In the video, a smiling Welcome sends me warm greetings, tells me I'm missed, and says he hopes I will return someday. It warmed and touched my heart to hear his words, see his smiling face, and know that he remembered me with fondness and appreciation. In the summer of 2006, I emailed the fund-raising manager at SANCCOB to send a donation, and in my postscript, I asked her to please pass along my greetings to Welcome. The next day I received an email thanking me for my donation, but her message ended with a few sentences that left me stunned and shaken to the core. She wrote that Welcome had unexpectedly died from AIDS in December 2005. I sat frozen in front of my computer, staring at the monitor in disbelief. Tears streamed down my face as I absorbed the loss of this truly special man, who had the most gentle and compassionate spirit. I cried because such a beautiful person had his life cut short in its prime; I cried knowing that he'd had to endure the ravages of a merciless disease that has killed so many in his country; I cried for his wife and his children, whose grief at having this sweet man ripped away from them prematurely must have been unbearable; I cried for all the penguins that would never benefit from his calming touch; and I cried because I would never have the chance to see him again or thank him for all he had done for so many thousands of oiled and injured penguins, and for touching my soul in the process.

16

Robbery at Gunpoint—Our Allegiance Is Tested

We were on a rescue mission, and to have members of our team violated like this was abhorrent. It took the wind from our sails, but we got up at 6 a.m. the next morning, rallied ourselves, and started the new day at Salt River . . . the penguins needed us.

—STEVE SARRO, AZA TEAM MEMBER FROM THE BALTIMORE ZOO

Saving the penguins became not just a physical challenge but an emotional one as well. In addition to struggling with the rigorous demands on our bodies and hearts, there came a time when some of our team members found themselves in a situation that was truly life-threatening. As it turned out, my parents' fears about encountering crime in Cape Town had been warranted, and had I been running just ten minutes late on the evening of July 9, 2000, my South African experience would have had a terrifying plot twist as well.

At the end of my dreadful first day at SANCCOB, I was supremely relieved when the shuttle driver arrived to pick me up. I was looking forward to getting back to the hotel and seeing my teammates again. The driver dropped me off at the hotel around 10 p.m. and I went into the restaurant to order some food. My colleagues were still at Salt River, so I sat alone in the dining room eating my supper. Eager to speak with everyone after our long day apart, I stayed at the table after finishing my meal to write in my journal and await the team's return. As I was recording the day's events, I suddenly remembered something important that I wanted to discuss with Jay and Linda, so I gathered my things and walked down the hallway to Jay's room. About ten minutes into our conversation, we heard a loud commotion outside and stepped into the hallway to see what was happening. Looking down the corridor, we saw that one of the glass doors to the restaurant was shattered, and several of our colleagues stood nearby looking dazed. The small group included Bonnie Boodram, Sharon Clouston, and Susanne Cloud (all from Earthkind in the UK), Bruce Adkins from the IBRRC, and Alex Waier from SeaWorld, who was a member of my immediate team. As we hurried over to speak with them, a frightening story unfolded.

Apparently, just after I left to speak with Jay and Linda, the five had returned from Salt River and gone into the restaurant to order some dinner. Moments later, two men rushed in, demanding everyone's money and valuables. One of the men was wielding a gun and the other was brandishing a large machete. Waving his gun in their faces, the first man said threateningly, "This is not a game. Put your hands up." The robbers then ordered them to lie facedown on the floor, and to keep their eyes closed. Terrified and shaking, all five did as they were told, and held their breath as the men rifled through their backpacks, looking for anything of value. Because the team had been at the rescue center all day, they weren't carrying much with them–only a few cameras and a small amount of cash, which the thieves took. As they lay with their faces pressed against the cold tile floor and their eyes squeezed shut, the man brandishing the gun continued to intimidate them by placing it against the backs of their heads and cocking the trigger.

One of the recipients of this treatment, Bruce Adkins of the IBRRC, had just arrived in Cape Town that very morning. This was definitely not how he had envisioned his first day in South Africa. As he recounted the frightening incident later, he said he was certain–when he felt the muzzle of the gun on the back of his neck and heard the trigger cock–that his life was about to end. Alex recalled thinking as he lay there next to the others, "I did not see your face, dude, so do not cap me because you're bored. I have a nephew I haven't even seen yet." After determining that they had few valuables on them, the two thieves emptied the cash register, stole $1,200 from the safe, and ran from the restaurant, smashing the glass door on their way out. Fortunately, all five of our colleagues escaped without being physically harmed.

As our teammates recounted their terrifying experience, I slowly took several deep breaths, realizing how close I had come to going through it with them. I could have easily been the sixth victim of this crime–just ten minutes earlier, I had been sitting in the restaurant where it occurred. The thought made me shudder. Had I fed one more penguin that day; had the shuttle to the hotel arrived ten minutes later; had I ordered something to eat that took longer to cook; had I not suddenly thought about discussing something important with Jay and Linda, I would have been lying on that restaurant floor beside my colleagues, silently praying for my life.

It quickly became clear to police and the hotel managers that the robbery was an inside job. The thieves had staged their attack precisely when the guards were changing shifts. As a result, no guards were actively on watch; the only person who could have been privy to this schedule and the accompanying protocols was a hotel employee. Concerned for our safety after this incident, the rescue directors decided to move everyone to a hotel downtown two days later, except for me and three others who were already based at SANCCOB–Sam Petersen, Karen Trendler, and Mike, a volunteer from the States. As SANCCOB was much closer to the hotel we were already staying at, it was decided that the four of us would not move with the rest of the team.

This did not sit very well with me. After all, if they believed it wasn't safe for the rest of the team to be at our current hotel, then shouldn't the four of us be moved, too? Sarah and Jay were quick to point out that our colleagues might not be entirely safe in their new downtown location, either. Although there was better security at the new hotel, our colleagues were warned that they should not go outside before 8 a.m. or after 5 p.m., as they would be likely targets for criminals. (As their long days began at dawn and ended after eleven, this was not going to be an issue–everyone was far too exhausted after returning from Salt River to even think about going anywhere.) But because the perpetrators of the robbery at our hotel and the person who tipped them off had not been caught, they still felt that moving most of the team was a safer option. Not only did the new hotel have enhanced security, it was much closer to Salt River, so the group would not have to travel as far to and from the rescue center each day.

It was very difficult to be separated from the rest of my immediate team at this point, as we had already formed strong bonds and I could have used their reassuring presence now more than ever. Staying at the hotel where the attack had occurred was unsettling, especially without most of my colleagues there. Our suite had an outside entrance, and the thieves had come right past our room on their way into the restaurant. It would have been remarkably easy for attackers to force their way in unseen, and now I would be staying in the room alone. To calm my nerves, I decided to use the same crazy logic that Garp employed in the movie *The World According to Garp*. Garp is pointing out a potential house to his new bride when a small airplane suddenly crashes into the home. When the dust settles, the tail of the plane is protruding from the second story; plywood, siding, and shingles are strewn across the yard. Viewing the destruction, Garp excitedly turns to his wife and tells her that this is the perfect house for them to buy because it has already had its disaster. Using Garp's reasoning, I chose to believe that this hotel had already had its disaster as well. Garp's unconventional approach is analogous to the old expression, "Lightning never strikes twice." (Except that this isn't really true–lightning can and *does* strike in the same place twice. I made a conscious decision to forget that I knew this as well.)

As would be expected, the victims of the robbery were deeply shaken by the event, and crisis counselors were brought in to speak with them the next day. All of the rescue team managers and supervisors working at Salt River were invited to participate in a group counseling session afterward as well. Although they had not been the victims of this frightening attack, everyone was under a tremendous amount of stress from the unrelenting rescue effort, and talking about their intense and trying experiences was cathartic for them all. Many tears were shed, and the bonds between team members grew even tighter as they spoke of how the robbery, and this enormous and harrowing animal crisis, had affected them. Because I had just been transferred to SANCCOB, I was unable to participate in this counseling session at Salt River with the others, which I deeply regretted. I really could have used a good cry and venting session, especially after the debacle of my first day at SANCCOB. Knowing I had missed out on connecting with the team in this way made me miss them even more.

Everyone was understandably concerned about their safety after this robbery, and during the counseling session, they discussed whether they should stay in South Africa or head home. The rescue directors made it clear that if anyone wanted to leave, they would not be faulted for doing so. Despite feeling somewhat uneasy, each member of the team ultimately decided that it was more important to stay and complete the mission. In the end, it really was not a difficult decision to make. There was a clear consensus amongst the group: they weren't going to let some thugs prevent them from doing everything they could to save the penguins. As Sharon Clouston, one of the robbery victims, resolutely told reporters, "This is not going to stop us doing the work we're here to do." It was settled; we all would stay.

One of the remarkable outcomes of this incident was the response of the local people we were working with. It seemed that everyone, from our team leaders to the volunteers, showered us with support and heartfelt apologies. In a statement issued to the press the day after the robbery, Sarah Scarth of IFAW South Africa expressed how appalled the agency was about this incident, saying, "Our team has been working very, very long hours and in exceedingly difficult conditions to bring help and assistance in this crisis. It's hugely disappointing that the very people who have come to South Africa to assist us in our time of need, have fallen foul of this country's crime epidemic." Dozens of volunteers approached us in the days following the robbery, telling us how very sorry they were that this had happened to members of our team. They clearly felt awful about it; we had come to their country to help them save their penguins and this was how we had been treated. They were both outraged and saddened that some of us had become victims of the violent crime wave in their city. It was as though each one felt some sense of responsibility, and they couldn't apologize enough. We were touched by their genuine concern and by their sense of remorse, and their kindness made us feel even better about our decision to stay and continue our important work.

While this incident made the local news, as well as news in the United Kingdom and the United States (since some of the victims were from these countries), it was not major network news and, thankfully, my parents never heard about it in the comfort of their living room back in Massachusetts. I know if they had, they would have been very concerned about my staying on in South Africa, and it would have put undue stress on them. My mother, especially (famous for worrying herself sick over every little thing) would have been in a state of full-blown panic had she known. Even after returning home to the States, I never told them about it. Knowing my mother, she just would have worried retroactively. In the ten years since the oil spill rescue, I never mentioned it to either one of them. My mother passed away six years ago, none the wiser. My father will learn of it now.

17

Leaving the Penguins Behind—A Difficult Departure

It was quite a hard transition at the end—it was a hard life. It was an incredible high, but it was long enough that when we returned to regular life, there was an element of a relief. But you miss the people, and it's always about the people. You miss giving the hugs, you miss the smiles, you miss the jokes. You made friends—you miss the people.

—MIKE HERBIG, SALT RIVER VOLUNTEER COORDINATOR

After eighteen punishing days, Heather and I had to return to the States. Most of our team would also leave this week, though Lauren would stay on another twelve days to finish raising the rest of the chicks. Our time in South Africa had been spent entirely in filthy rehabilitation centers, interspersed with brief intervals of much needed sleep in our hotel rooms. All we had seen of the country were the interiors of dark, dusty buildings and thousands of oiled penguins. Due to the all-consuming nature of our work, we never had a spare moment to take photographs, so on our way out of Cape Town that last morning we stopped by Salt River to take some pictures and say our goodbyes. On our final walk through the mammoth rescue center, I hurriedly snapped a few rolls of film, trying to capture the size of the operation and the environment as best as I could. But the lighting was poor and there wasn't time to compose great photographs, so these hastily taken snapshots—plus the dozens of scars on my arms and hands—would have to suffice as my only permanent record of our experience. (When we returned home without one picture of either of us in the rescue centers, our boss, Dan, was beside himself. He was incredulous that we had been in South Africa that long, and for such an important mission, yet had not once thought to document our presence there. Although we had just taken part in the largest live animal rescue in history, it never occurred to either one of us to get pictures of *ourselves.* We had far more important things on our minds!)

When I was last at Salt River, ten days earlier, all of the penguins were still housed inside. Now, many had been washed and moved to the newly built outdoor holding area; it was good to see some of the penguins outside in the sun in large, open, less confining pens. After watching them enjoy a swim in one of the large outdoor pools, we went back inside the warehouse and made our way from one enormous room to the next, sharing hugs and tearful goodbyes with our colleagues, team leaders, and volunteers, all of whom we had grown close to in our short time there. (When Lauren later told me that, on her last day at Salt River, she felt just like Dorothy saying goodbye to her newfound friends at the end of *The Wizard of Oz,* I knew exactly what she meant.) We thanked the volunteers again for everything they had done to help us, then climbed up onto a raised platform overlooking our room to take in the mind-boggling scene one last time.

As we looked out over the vast interior of the building, we could clearly see that there were still thousands of oiled penguins waiting to be washed and rehabilitated. Incredibly, 7,000 penguins had already been cleaned in the short time we had been there—but they hadn't finished undergoing the process of being rehabilitated, and had yet to be released. At this stage, we did not even know if the penguins would be successfully returned to the wild. It was agonizing to leave when there was still so much work to be done, and I felt horribly guilty abandoning our colleagues and the penguins, but we had our own penguin colony to care for back at the New England Aquarium. At the pleading of the rescue directors, we had already stayed in Cape Town nearly a week longer than originally planned, but it was the middle of the breeding season at the aquarium and our help was needed there to raise several newborn penguin chicks.

After leaving Salt River, we made a quick stop at SANCCOB. Heather had done some work with SANCCOB a few years earlier when she traveled to South Africa to help with the African penguin census, and she wanted to swing by to see the folks she had met during her last trip there. I also wanted to say my goodbyes to the wonderful people I had come to know during my nine days at the center. I filmed some hasty videos, and took several photographs of staff and volunteers feeding and swimming the birds and cleaning their overcrowded pens. As I passed by the small administrative office inside, Kumie Joubert stuck her head out to greet me. She was in her usual uniform: long, black wool coat and white sneakers. Short and pleasingly plump, Kumie was one of the few people manning the phone lines at SANCCOB during the rescue. With hundreds of calls coming in every day from people eager to volunteer, this alone was a full-time job. Kumie had an easy smile and she typified the thoughtful and pleasant nature of almost every Cape Town citizen we met. She knew I would be stopping by that morning, and said she had something for me. She handed me a manila envelope; inside were several printed pages of photos with inspirational quotes that she had put together for me. She had picked out quotes that spoke eloquently to the challenging yet rewarding situation we had found ourselves together in. I wept openly as I read the carefully chosen words–they truly touched my heart, and I was deeply moved by Kumie's sincere gesture. After more tears and hugs with the rest of the folks there, we climbed back into our rental car and headed off.

Because the penguins that had been under our care were still being rehabilitated and we would not see a release, we decided to drive to Boulders Beach on our way to the airport to see some healthy penguins in their natural surroundings. Located about forty-five minutes south of Cape Town, Boulders Beach is the location of one of the few mainland breeding colonies for the African penguin. Established by the penguins in 1982, it has grown to about 3,000 individuals since then. On this picturesque beach that is also popular with locals and tourists for swimming and sunbathing, the penguins are often seen nonchalantly waddling across beach blankets on their way to the water. They've also been the source of frustration for some local residents due to their habit of nesting in flower gardens or wandering uninvited into their homes. And their loud, donkeylike braying at all hours of the day and night doesn't do much to boost their popularity with the locals, either. These mainland penguins have clearly become accustomed to being in close proximity to humans, though they won't hesitate to bite someone when cornered or provoked. With the help of local conservation groups–and a substantial donation from Pick 'n Pay market–their nesting grounds have been cordoned off in an area of Boulders called Foxy Beach, and raised walkways have been constructed so tourists can view the penguins without disturbing them.

After arriving at the beach and parking next to a green sign that cautioned us in bold white letters WARNING–PLEASE LOOK UNDER YOUR VEHICLE FOR PENGUINS, we started walking along the meandering wooden boardwalk that snaked its way to the waterfront. This would be my first time viewing African penguins in the wild, and I was very anxious to see birds that weren't covered with oil, starving, or traumatized. The Boulders penguins had not been seriously impacted by the *Treasure* oil spill because their colony was on the eastern side of the Cape Peninsula, while the oil spill had occurred on the western side of the same finger of land. As we approached the rookery, my excitement mounted: I could hear the penguins braying in the distance, and quickened my pace, eager to finally see them in their natural habitat. When at last we reached the viewing platform, we stood there for a long time in silence, taking in the scene. After being surrounded by oiled, sickly penguins for so long it gave us a deep spiritual lift to see clean, healthy, active penguins swimming in the surf and basking in the sun with fat, fluffy brown chicks by their sides. There were penguins digging nests in the sand, penguins braying and displaying for their mates, parents feeding large chicks, and others happily preening.

After watching for a while and taking several photographs of clean penguins and their robust chicks, we walked across the street to have lunch at the Boulders Beach Lodge Restaurant. Having eaten very little for the last eighteen days, I slowly savored every tangy bite of my mango-chicken croissant sandwich, while gazing at the sparkling ocean and dramatic, rocky coastline through the huge glass windows at the front of the restaurant. It was a breathtaking vista. Before that day, I'd had no idea that there were places in South Africa that looked like this. I took a mental snapshot of the jagged brown cliffs dropping down to meet the expansive azure sea–the only bit of South Africa I had seen outside of SANCCOB, our hotel, and the Salt River warehouse. I thought to myself, "I have to come back someday and actually see this beautiful country." But at least I had seen and experienced its gracious and caring people, and they would remain in my heart always. After finishing our lunch and buying a few small souvenirs at the restaurant, we drove to the airport and boarded the plane that would take us back to America, our lives changed forever by our time in South Africa.

The transition back to life at home and the daily routine at work was jarring and unexpectedly difficult. We had spent eighteen solid days working non-stop, day and night, in a dark, filthy warehouse, surrounded by thousands of oil-soaked penguins, with one objective in mind–to save the afflicted birds and return them to the wild. The experience in Cape Town was intense and surreal, and I had been in an altered mental and physical state during my time there. I had worked exhausting sixteen-hour days, had barely slept or eaten, and had been under more stress than I thought humanly possible. South Africa was a full-immersion experience. And when I was abruptly wrenched from that intense, relentless existence and thrust back into my former life, I felt as though the ground had dropped away from beneath my feet. The return to life as I had known it before was extremely disorienting, and it took several weeks before I could get my bearings. It seemed there were two separate and opposing realities that were at great odds with each other: there was the penguin rescue, which felt vivid, immediate, and real; and then there was this other life I'd had before going to South Africa, which felt strangely distant and nebulous.

This confusion even plagued my dreams, playing tricks with my subconscious. The morning after our return, I awoke to the sight of thousands of oiled penguins filling every corner of my apartment. I was instantly in full readiness mode, but this time there was no one there to help me take care of the birds. I was momentarily puzzled by the presence of all these penguins in my bedroom, and wasn't quite sure if I was at home or at Salt River. But, knowing I was alone, my immediate reaction was panic and dread, as I began the usual mental gymnastics of figuring out how I was going to feed, clean, swim, and rehabilitate them all. Eventually, I woke up enough to register that I was no longer at the rescue center, and that the throngs of oil-soaked penguins crowding my apartment were merely part of a waking dream. Once that realization set in, my mind stopped racing and my galloping heart slowly returned to its normal rhythm.

This unsettling experience repeated itself every morning for more than a week after my return home, and each time I woke in a state of alarm before realizing the penguins weren't really there. My teammate Alex also suffered from these waking nightmares. Ten years after the rescue, there are still some mornings when he is tormented by them. "I occasionally wake up with this *'How are we going to get all this done today?'* panic," he told me. "I never had that happen before the oil spill rescue." The disturbing dreams that many of us had after the rescue reminded me of the anxiety dreams I used to have when I waited tables. These dreams were always the same: I'd be the only waitress on duty and the restaurant would be empty. One party would come in, sit in the corner, and I'd go over to take their order. When I'd nonchalantly turn back around, the entire restaurant would be completely packed with people and I'd be instantly in the weeds! Although the penguin dreams eventually stopped, it took longer for the bone-numbing exhaustion that had plagued me in Cape Town to dissipate. The twelve pounds I had lost while there, however, returned fairly quickly.

It was nearly impossible to describe to my friends, family, and colleagues at the aquarium what I–and the penguins–had been through, as there was nothing in most people's experience to which it compared. There didn't even seem to be the proper vocabulary to describe such an extraordinary event and the powerful emotions accompanying it, and I felt that the only people I could truly relate to were others who had been there. Before we left South Africa, Jay, Linda, and Sarah had told us this was not uncommon, and that talking with other people we had shared the experience with would help us work through the unique and complicated emotions we were bound to have. They also warned us that people who worked unrelenting wildlife rescues like this one often experience a mild form of post-traumatic stress disorder, including disturbing dreams and inexplicable mood swings.

My teammate Lauren DuBois was profoundly affected by her experiences in South Africa and struggled to reacclimate after returning home to California. She found it painful having nobody who could relate to what she had gone through, and, like me, she found it impossible to articulate exactly what the rescue effort felt like, smelled like, looked like, *was* like–and how it had affected her. Only someone who had lived through it could truly understand it. This inability to share the experience with the people in her life left her feeling lonely and disconnected, and without warning or provocation, she'd become frustrated and snippy with her co-workers. "How can they complain about work when they have it so bloody easy?" she'd think. "Try sitting in a room surrounded by seven hundred chicks that need to be fed, knowing you can't possibly feed them all that day!"

And while she found the return to "normal" life challenging on many levels, it was some of her duties as a chick-rearing supervisor that left her struggling to reconcile the tough choices she'd had to make while in South Africa. The morbid responsibility of determining which tiny chicks would live and which would die had left an indelible mark on her psyche. Lauren was also haunted by bizarre dreams, but hers were far more disturbing than mine had been. Her recurring dreams featured cardboard boxes of baby penguins, their heads still crowned with brown downy fluff, being flushed down the toilet one by one. As they swirled down the drain, the chicks looked up at her imploringly, and Lauren could only watch helplessly as they disappeared from her sight. Her vivid dreams about these chicks were clearly evocative of the heart-wrenching and impossible choices she'd had to make during her time in South Africa.

One of the other challenges I soon encountered after returning home was finding significant meaning in everyday life. The work we had been doing in South Africa was vital in saving the lives of thousands of animals–and potentially the future of an entire species. On the heels of that experience, commuting to work and going through my daily routine felt somewhat hollow and pointless. While I knew, intellectually, that the animals needed me, and that educating the public about pressing conservation issues was important, this seemed to pale in comparison to the work we had been doing at Salt River and SANCCOB. I struggled to find joy and meaning in the work I had been doing before going to South Africa; but for a while, I just went through the motions, waiting to feel a sense of purpose and enthusiasm again. As I drifted aimlessly through each day, I began to wonder if I would ever feel the same about my job and my life again. It would be some time before a sense of normalcy returned, along with the dedication and passion for my work I had previously enjoyed.

Part of the difficulty in adjusting to our lives back in America was the lingering knowledge that our colleagues were still elbow-deep in penguins at the rescue centers in Cape Town, and we were unable to help them. The guilt was overwhelming at times. We'd had to leave while there were still thousands of penguins to clean, rehabilitate, and release. I couldn't help thinking that the emotions I was experiencing were similar to survivor's guilt. After responding to this dire animal crisis, we'd deserted the others when they needed us most. We had survived the horrific reality that was Salt River, but our colleagues were still in the thick of it. Even though we were stressed out and exhausted, we'd wanted to see it through to the end. Our greatest wish was to see every last penguin washed, restored to health, and set free. But we'd left before even one penguin had been returned to the wild.

There was a sense of incompletion; of not fulfilling an important obligation; of not doing everything humanly possible to ensure these birds would survive. Because we left so early in the rescue, not only were we plagued by feelings of guilt, but we did not have a sense of closure. Leaving when things were still so uncertain was difficult. So many questions hung in the air: Would the penguins survive? Would they all get cleaned and returned to the wild? Would enough volunteers show up to help care for them? The feeling that we could and should be doing more would not leave us for some time. When we flew out of Cape Town, we had no idea how the penguins would fare. The majority of the birds had a long way to go before they'd be ready for release, and there was still a tremendous amount of work to be done before their fate would be known—not only for each individual bird but for the entire species. Would the unwavering efforts of more than 12,500 caring people make a difference for the penguins? Or would all of their hard work have been for naught? Now, along with the rest of the world, we could only watch from afar and hope for the best. We all would have to wait to learn their destiny.

18

Hope on the Horizon—The Penguins Survive and Thrive

The massive penguin rescue and rehabilitation effort under way is unprecedented. The outpouring of international public support has been wonderful and has meant that this mission can continue. Whether it is volunteers coming forward or international donations being made, the message is clear—the world wants to save these unique birds.

—SIMON POPE, IFAW-SA

The mammoth rescue effort in South Africa was never far from our minds. The work at the rescue centers was still arduous and unremitting when we left Cape Town, although by this point defined systems and a slightly more manageable routine had been established. Rotating teams of penguin specialists continued to fly in every few weeks to help rehabilitate the penguins and supervise the volunteers. We looked forward to periodic email updates from the friends we had made while in South Africa, telling us how things were going. Even if we couldn't be there ourselves, it was nice to be kept abreast of what was happening with the penguins, and we were heartened to hear about the progress they were making each day. Although many of the regular volunteers were now well trained and working independently, school was back in session and the number of volunteers at the two rescue centers had fallen steeply, adding a new element of stress to the operation. Finding enough people to help care for the birds became a serious challenge and a constant source of worry. As the rescue effort entered its second month, and then its third, the number of volunteers kept dropping. At the beginning of the operation, up to 1,000 local citizens rushed to Salt River each day; now, they were lucky if 300 showed up. On their worst day, only 72 people turned out to help. The rescue effort was dragging on for so long that some people probably grew tired of hearing about it, and perhaps didn't feel the same sense of urgency they had felt at the beginning. But the penguins still needed them as much as they had on the first day of the oil spill.

The emails that came from people I had met at SANCCOB and Salt River emphasized the continuing struggle to keep the rehabilitation efforts moving forward. After we had been home for five days, I received an email from Patrick Maloney, one of the many dedicated volunteers I worked with. "There has been a drop in the number of volunteers coming in," he informed me, "but not enough to lower the spirits!! We washed 635 birds on Wednesday–a new high! We had to catch up on birds not washed a few days before due to a lack of volunteers. Everyone just put their heads down . . . scrub, scrub, scrub. Over 600 sparkly clean birds in one day! It was quite an emotional thing!" He went on to describe how Betty, one of the workers from Florida, had passed out from sheer exhaustion over the weekend. She had been running one of the washrooms at Salt River, but rescue directors put her on an enforced two-day break after her collapse. "But she will be back," Patrick continued. "She, and her colleague Kim, are awesome people, like every single other person volunteering!" As we had seen during our time in Cape Town, everyone involved in this rescue effort had an indomitable spirit, and nothing could keep the staff or the volunteers from fulfilling their important mission.

Gerald Van Wijk, another unswerving volunteer in our room at

Salt River, also kept me posted on what was happening at the rescue station. He and his two close friends, Rick Meijer and Roland Jolink, were among the group of tremendously devoted volunteers who helped keep us on an even keel every day in Room 2. Always smiling and endlessly helpful, "The Three Musketeers" (as I called them) embodied the warm and welcoming spirit of all the South Africans we met. Rick and Roland owned Black Heath Lodge, a B&B in Cape Town, and they generously put up anyone volunteering with the rescue for half price. And Gerald arrived at Salt River one day bearing bars of beautifully scented organic soap for Heather and me. Wrapped in handmade paper and tied with twine, it smelled of lemon and lavender. After enduring the stench and filth of that warehouse for sixteen hours a day, this was truly a very thoughtful gift. He told us that it was "metaphysical soap–to wash away the negative emotions." I used that soap every night for the rest of our stay.

In an email sent shortly after we left Cape Town, Gerald lamented the dreadful cost of oil spills on wildlife, citing another deadly event that was occurring at the same time. Shortly before we left for home, news of a catastrophic spill in Brazil had reached us in South Africa. On July 16, more than 1 million gallons of crude oil had poured into the Iguazú River near Curitiba after a pipe burst at a refinery owned by the oil giant Petrobras. The company had an appalling record of major oil spills, but this was the worst the country had seen in twenty-five years. The few Brazilian veterinarians and penguin specialists we were working with at Salt River were suddenly faced with an agonizing decision: Should they stay in Cape Town to help save the 19,000 oiled penguins already at the rescue centers? Or should they go back to Brazil to help save the animals now trapped and dying in the oil spill there? Though our colleagues were torn, they knew they had to return to their home country to help deal with the fallout from the devastating environmental disaster taking place there.

"Our Brazilians have left," Gerald reported, "as they now have their own oil spill to contend with! Unbelievable, when we have just had one of such magnitude . . . does no one stop and realize what the sea life has to go through? How do ships still sink in this day and age? This is all very new to me, so I can imagine how discouraged some of the people feel who do this all the time. Or does the compassion for the animals make it all worthwhile?" I have often wondered the same thing myself: How do Jay, Linda, Sarah, and Estelle–and all of the other amazing people whose life mission it is to rescue injured and oiled wildlife–manage to do this type of work day after day, year after year, decade after decade? How do they care for thousands of oiled, injured, and dying animals for years on end without falling into a deep pit of despair, hopelessness, and anger? I don't know how they are able to keep it all from getting to them. I deeply love animals, and to see them suffering is just too much to bear at times. Somehow, the tremendous reverence and concern that Jay, Estelle, and the others have for these helpless creatures must force them to push through the pain, allowing them to focus on the vital work of healing the animals and returning them to the wild. Their selfless and compassionate efforts are truly remarkable; it's just shameful that mankind has caused so much of the devastation that makes their work necessary. They are a special breed, these people, and true heroes in this world.

Despite the loss of the Brazilians and hundreds of volunteers, the hard work at Salt River had to go on. One of the goals of the rehabilitation team was to get all of the penguins in the Salt River warehouse outside as soon as possible, as this would reduce the likelihood of any airborne illnesses spreading, and lessen the stress on the penguins from being crowded into the relatively small holding pools. While we were still in Cape Town, large enclosures were being constructed outside, and some of the birds had been moved to these areas. Shortly after we left, the rest of the penguins were moved outdoors as well, where they could enjoy the benefits of sunshine and fresh air. Several tons of shale and sand were trucked in to provide an appropriate surface for the penguins to stand on, and additional outdoor pools were built for their daily swimming sessions. Aside from rinsing the guano off their feathers and helping to regain their waterproofing for a safe release, these enforced swims also got the penguins off their feet for a while, which helped reduce the risk of getting a foot infection called bumblefoot, and gave the birds a chance to exercise.

A few penguins, though, had their own clandestine plans for an early release. Apparently eager to return to their island homes, these cunning birds were determined to escape and find their way back to the ocean. After all of the penguins had been moved into the large holding pens outside the Salt River warehouse, a few wily birds managed to slip out of their enclosures. One was found waddling down the train tracks that ran from Salt River toward the coast. It was actually headed in the right direction, and was caught only a mile away from the ocean. The penguin just might have made it to the water had it not been spotted by an alert citizen, who bravely captured it and returned it to the warehouse. Wrapped up inside the woman's jacket, the thrashing, kicking penguin was none too happy about being detained, but it was not fully rehabilitated yet, so it was fortunate to have been discovered before it reached the ocean.

Another motivated penguin at Salt River had its own adventure. Having escaped its pen, the bird managed to make its way to the city center, more than a mile away. Calls began coming in to the police station about a loose penguin, and an officer was dispatched to retrieve the runaway bird. Having nowhere else to stow the penguin after finding it, he put it into the holding area of his vehicle. But before he could bring the bird back to the rescue center, he got a call about a robbery in progress. He apprehended the criminal and put him in the back of his van, along with the intercepted penguin. The two, no doubt, kept their distance and eyed each other a bit nervously. After stopping at the police station to drop off the man he had in custody, the officer drove down to Salt River and handed over the fugitive penguin to the rescue team.

By the time the penguins had been moved outside, most of them had made the transition from being force-fed to voluntarily taking fish from people's hands the way penguins at zoos or aquariums will. The relative ease and speed with which they learned to eat this way was remarkable, and a testament to the intelligence and adaptability of these resilient seabirds. This method of free-feeding greatly reduced the time required to feed all of the penguins, and it also lessened the stress on both the handlers and the birds. Getting a wild penguin to eat from your hand is not quite as simple, though, as standing beside a holding pen, offering a fish, and having the penguin waddle over and eagerly slurp it down. While a handful of penguins grasped this concept without too much coaxing, the majority of the birds had to go through a number of steps to transition from being force-fed to free-feeding. To train the penguins, feeding boxes were introduced. Mariette had first invented these feeding boxes during the *Apollo Sea* rescue effort, and found they were a very effective training device.

Rectangular in shape, each feeding box was approximately one foot deep by three feet wide by two feet high. The top was left open, and on each long side of the box were doors that slid up and down like a guillotine. To fashion the new training system, two of the penguins' holding pools were placed next to each other so that they were touching. One was full of penguins; the other was empty. The round pools were constructed of long sheets of vinyl wrapped around wire cages, so the ends could either meet (forming a complete circle) or they could be left open, leaving a gap. A gap of three feet was left open in both pools where the two pools met, and one of the feeding boxes was placed in the open gap, so that the two pools were then connected by the box, forming a figure-8 shape. Once this structure was set up and the feeding box was in place, the training of the penguins could commence.

The sliding door on the wooden box facing the penguins was lifted, and four or five penguins were herded into the box. The door was then slid down, separating those few penguins from their pen mates. Once the penguins were in the box, they were offered fish without being grabbed and force-fed. With the birds confined to this small space, and with no way to escape, most eventually opened their mouths to eat when a pilchard was firmly and persistently pressed against the sides of their beaks. Once each individual bird had eaten its fill, the sliding door was lifted, and the bird was allowed to escape into the clean, empty pool on the other side. Each penguin was kept inside the feeding box until it had eaten on its own; if a penguin still refused to eat after all the other penguins in the box had eaten, then it was force-fed. It typically took a few days before the penguins figured out what they were supposed to do. At first, many birds still had to have their beaks pried open and the fish placed into their throats; but eventually most of the penguins caught on and, once inside the feeding box, looked toward their human caretakers, beaks open wide, waiting for the fish they now realized was coming.

Within a week, most of the birds had become accustomed to free-feeding inside the boxes. After each pool of penguins had been fed, if there were enough volunteers available, the birds were given additional opportunities to eat throughout the day. A feeder would stand by the side of each pool offering fish, while the penguins inside voluntarily approached and opened up their mouths to be fed. At first, many of the birds were quite tentative; but after several days, the whole group of birds inside each pool would push and shove their way toward the feeder, mouths agape, greedily lunging for the fish being offered; some would even yank pilchards out of other penguins' mouths and gulp them down. Often, after feeding from outside a pool, the volunteer would then sit inside the pool to offer more fish to the birds. There usually were several holdouts in each pen; these slow learners were more leery of us, and refused to approach the feeder to take any fish. After being grabbed, restrained, and force-fed every day, it was easy to understand their nervousness. A hungry penguin, though, is a motivated penguin and, for the most part, hunger eventually overrode any fear they had of humans.

When you stop and think about it, it truly is remarkable that a wild animal could adapt so quickly and be trained so easily. I don't imagine there are many other untamed animals on earth that could learn not only how to eat directly from people's hands but also to approach voluntarily to be fed. Penguins are very bright birds! Some of the adult penguins became so accustomed to certain people that their behavior could be quite surprising. Mariette was being interviewed one day for *Kwela,* an Afrikaans television program. She had been working with a particular pen of penguins for two months at that point, and the birds had become used to her presence. As she sat on the floor in their pen during the interview, two of the penguins surprised her by waddling over and checking her out very closely. After circling her a few times, one of them climbed up onto her lap and just sat there, hanging out. They weren't begging for food, as they had recently been fed–they just seemed to want some attention. She found it quite remarkable that these wild birds would behave in this way, and was flattered that they were so comfortable with her.

An interesting phenomenon was noted during the process of training the penguins to eat from a person's hand. The penguins that first made the transition to free-feeding were those that had been oiled during the *Apollo Sea* oil spill six years earlier; they had been through this stressful rehabilitation process at least once before. These *Apollo Sea* birds were identifiable by the metal ID bands they wore on their wings, souvenirs from their previous rehabilitation experience. (In fact, one of the two penguins that checked out Mariette during her interview was a bird she had banded during that earlier rescue.) Something in their brains clicked, and they recalled that this strange way of taking food was much easier than getting manhandled every day and having pilchards shoved down their throats. Because animals often learn through imitation, most of the "rehabilitation virgins" picked up the technique of free-feeding from watching their bolder and more experienced counterparts. Being very resourceful birds, it didn't take long for most of the penguins to figure out the new routine. Once they had made this transition, the entire feeding process was much faster and safer for the worn-out volunteers, and easier on the stressed-out birds as well.

The primary goal of the rehabilitation process was getting the birds back into the wild as quickly as possible. Prior to being released, however, each of the 19,000 penguins had to undergo a number of critical tests and examinations. Blood was drawn to check their hydration and iron levels; they were given a series of vitamin injections and immunizations; and they were weighed to be sure they were robust enough to survive if food at sea was scarce. Also, before being released, each penguin had a metal ID band placed on its wing, which would allow researchers to monitor them in the future. Most important, each penguin had to be "graded"–a long, tedious process in which each bird was forced to swim in cold water for an hour, after which a handler held the bird in their lap and ran their fingers backward against all of the feathers, lifting them to ensure that none of the downy base was wet. This meticulous procedure took about fifteen minutes for each bird, as every inch of their body had to be carefully examined. If any portion of the down underneath the feather surface was damp, it was a sign that the penguin had not completed the process of re-waterproofing its feathers. In this case, the bird in question was held back until its feathers were completely watertight. Releasing a penguin before it could remain completely dry in the water would have been a death sentence; if unable to enter the water to hunt they would eventually starve. Once they passed this final test, a pink dot was sprayed on their chest with a temporary paint that would last for several months. This helped to identify which penguins were ready for release, and it would also make it easier for researchers to spot the rehabilitated birds once they had returned to their islands.

Prior to being graded, each penguin was swum for a week or two after being washed, so it could rinse the last of the residual soap from its feathers and build up its waterproofing again. The enormous pools that had been built outside Salt River provided a setting where large groups of penguins could swim in an environment that was slightly more natural than the small, shallow swimming runs they'd been using while housed inside the warehouse. One afternoon, Big Mike took a rare break to watch the penguins as they bathed and preened in one of these large pools. "I remember walking by one day and watching penguins swim," he recalled, "and I spent a half hour completely mesmerized, watching underwater turbo-jetted aeroplanes. It was the most beautiful, elegant thing to watch. And I remember smiling, my head in my hands, just watching these jet-propelled engines underwater. It was kind of a step up from washing penguins, which represented hope. Here, we were actually achieving something. These penguins were just enjoying the incredible freedom of not standing on mats, not standing on land, but actually swimming. This was their habitat. Seeing these guys swimming and realizing: they're going back. That was a beautiful moment for me."

Another key event that had to take place before the birds could be returned to the wild was cleaning the remaining oil from their habitat. It would have been foolhardy to release the penguins before the rest of the oil had been removed from their islands and the sunken ship, as they would likely just get oiled again. In fact, one of the volunteers at Salt River put up a poster they had made, spoofing this potential disaster. On it, they had drawn a feedback loop indicating that, after the oiled penguins had been cleaned, they would be released, then they would swim right back through the oil still at sea, and return to Salt River again for a second washing–and so on, and so on. When I first saw the small poster taped to a wall by the entrance of Salt River, I didn't know whether to laugh or cry, as it was just too close to the truth. At that point in the rescue effort, nobody knew how much longer it would take to complete the oil remediation. Rescue managers were even starting to muse in jest which would be worse for the penguins: staying cooped up in stressful conditions in the crowded warehouse, or being set free and possibly getting oiled again, prompting a second stay at the Salt River Penguin Crisis Centre. Fortunately, they would soon get the news they had been waiting for.

The day after we returned home, a jubilant email arrived from our South African colleagues, informing us that the first group of rescued penguins had been released back into the wild. We had not been aware that on the very day we were flying out of Cape Town, this landmark moment was occurring in Milnerton, a suburb north of the city. On this same day–July 18–the last of the oil was removed from the sunken ship, and the coasts of Robben and Dassen Islands were cleared of oil contamination. This first release of rehabilitated penguins was a highly anticipated event, and it generated a great deal of attention from both the media and the general public.

A large crowd of reporters, rescue workers, volunteers, and Cape Town residents gathered to watch as boxes containing 262 penguins (59 from SANCCOB and 203 from Salt River) were opened on the beach and clean penguins streamed out. As the birds raced toward the breaking surf, the rescue team celebrated this symbolic moment that marked a notable turning point in their mission. When it began, the rehabilitation of 19,000 oiled penguins seemed to be a nearly impossible task. Now, witnessing the first release of cleaned penguins, there was renewed hope that the remaining penguins would also survive their traumatic ordeal. It was a much needed morale boost for the exhausted rescue workers, who were able to return to the rehabilitation center holding on to visions of clean, healthy penguins returning to their home in the sea.

For us, back in the comfort of our homes in the United States, it was a very poignant moment. It had been almost unbearable to leave before seeing any of the birds returned to the wild, as their fate was still unknown. And though I desperately wished we could have been there to witness the release of these clean, healthy penguins that we had helped to rehabilitate, the knowledge that this first group of birds had survived and made it back to the sea was deeply rewarding, and provided evidence that our contributions to the effort had been both effective and worthwhile.

19

Freedom for the Remaining Penguins—The Rescue Draws to a Close

Oh this has been a massive effort. It's been a huge effort. This is the biggest operation internationally ever in saving any species. I want to say that we are working on a miracle here. I [would] have never ever thought, when I realized how big this spill was going to be, that we were going to get the support we're getting, that the penguins were going to be getting the support they're getting, and that people were going to be so absolutely wonderful and ready to give in such a big way. I think we can all learn from this and we can all begin to have more faith in . . . in mankind. It has been absolutely wonderful. Thank you.

—ESTELLE VAN DER MERWE, AT A PRESS CONFERENCE

When the *Treasure* first sank on June 23, 2000, and rescuers began capturing oiled penguins to be rehabilitated on the mainland, there was a growing sense that this was going to be a huge event. As more and more penguins started straggling ashore on their breeding islands with thick, black oil dripping from their bodies, it became evident that this rescue effort was going to be unlike any other that had ever come before it. It wasn't just going to be huge–it was going to be truly monstrous. When SANCCOB surpassed its maximum holding capacity in just three days and then the Salt River warehouse began filling with oiled penguins, the reality and gravity of the situation became inescapable. The sheer number of birds that would need to be fed, swum, washed, and rehabilitated was astronomical; and for Estelle and the others charged with managing this crisis, feelings of shock, and then dread and doubt, started to creep in. They realized that they were facing the largest rescue of oiled seabirds ever undertaken, and the amount of work that lay ahead was unimaginable. It hardly seemed possible to save so many animals–it simply had never been done before.

In what had been the largest rescue of oiled penguins six years earlier, just half of the 10,000 birds taken in for rehabilitation survived. This time, not only were twice as many oiled penguins undergoing rehabilitation, there were also four times as many penguins being handled and relocated overall. It hardly seemed possible, given the enormous size, that this rescue effort could be any more successful than that smaller one had been. Anton Wolfaardt, the reserve manager from Dassen Island, noted that everyone was very concerned in the beginning. "When we started appreciating the magnitude of the disaster, it was certainly very worrying," he told me. "I knew that we had the experience and plans in place to deal with an oil spill–we had learned a lot of lessons from the *Apollo Sea*–but the *Treasure* was so much bigger than anything we had previously experienced."

As each of us stepped through the vast, yawning doors of Salt River for the first time, and our eyes fell upon the endless rows of blue pools containing thousands and thousands of oiled penguins, we struggled to absorb the horrific scene before us. Our brains reeled as we tried to imagine the amount of work required to save all of these birds and get them back out into the wild again. The whole thing was too massive and overwhelming to fully comprehend. Yet a positive energy permeated Salt River, as well as a tremendous spirit of compassion and cooperation, and a firm resolve on the part of each person to do everything in their power to help the traumatized birds. Even though the number of oiled penguins requiring rehabilitation was unprecedented, our team leaders were mindful to convey the attitude that we would be successful. They not only managed the mind-boggling logistics of the huge operation with military precision, they also led the wildlife rescue teams with confidence and optimism, which made us all feel more assured and certain that we *could* do this, that we *would* save these 19,000 penguins. Somehow, in the midst of chaos and disaster, we managed to have a sense of hope. This was also due, of course, to the incredible show of support by the extraordinary volunteers who came to help. Had the people of Cape Town not responded by the thousands to care for the penguins in their darkest hour, all hope would have been lost.

Now, more than three months after the first oil-soaked penguin had been rescued from Robben Island, the enormous effort was drawing to a close. Although a small number of birds requiring further treatment still remained at SANCCOB, the majority had already been released. They had come through the doors of Salt River and SANCCOB packed in cardboard boxes, weak and confused, their feathers matted and soaked through with heavy bunker oil. They arrived dehydrated, in shock, ill, anemic, and starving–and some were blinded from the oil or otherwise injured. They were wary of us, and they kicked, struggled, and fought their way through every step of their rehabilitation. But, finally, almost all of the oiled penguins brought to SANCCOB and Salt River were rehabilitated and set free. A few lucky survivors were sprung from the rescue center after just twenty-five days. For most, their forced confinement lasted about two months. A small number of severely compromised birds languished at SANCCOB for nearly three and a half months.

Slowly but surely, every last penguin had the viscous oil removed from its body. Each one had to endure the long, painstaking scrubbing until its bathwater ran clear, and its feathers were finally pristine and lustrous again. For a period of four solid weeks, staff and volunteers spent strenuous, sixteen-hour days bent over washbasins cleaning filthy, ill-tempered penguins. Finally, on July 29, just after 8 p.m., the long-awaited moment arrived. The last oiled penguin had been washed! A huge cheer erupted in the washroom as the bird was toweled off and brought to the drying room, where it was put under infrared heat lamps to dry out overnight. A bottle of champagne was opened, and the exhausted staff and volunteers shared a celebratory toast as the sparkling beverage was doled out in paper cups.

The average number of penguins being washed each day at that point was 550; but on this evening, the penguin washers pushed themselves harder so they could get through the remaining oiled birds. They were so close to completing their long task that they didn't want to stop, not when the end was so tantalizingly near. That Saturday, they washed 693 penguins. It was the second largest number of birds to be washed in a single day during the *Treasure* rescue effort. Incredibly, it had taken just twenty-nine days to clean all the oiled penguins at Salt River. On average, the washing teams cleaned 475 penguins a day. At the smaller SANCCOB facility, approximately 30 penguins were washed each day over a period of thirty-nine days. When all was said and done, nearly 17,000 penguins had endured the long, stressful washing procedure; and hundreds of volunteers had braved thousands of vicious bites in the process. Because of their hard work, these penguins now had a second chance at life.

Once they were washed and had regained their waterproofing, the penguins were gradually released in large batches. Following the first release of 262 rehabilitated penguins on July 18, periodic releases occurred approximately every other day through August 4; then they took place almost every day until August 24. The smallest group to be set free again consisted of just 72 penguins (on July 24) and the largest contained 1,352 penguins (on August 17). On average, about 550 penguins were released each time. When the doors of Salt River closed for the final time on August 24, another 85 penguins were released, and the remaining birds still in need of care (about 1,000 of them) were moved to SANCCOB, where they joined another 1,000 already being cared for at that facility. Nineteen thousand oiled penguins had passed through the doors of the two rescue centers since June 24; a hundred and nine days later, on October 10, the last victims of the *Treasure* oil spill were returned, at long last, to their island homes off the coast.

In addition to freeing the adults, rescue workers released the abandoned chicks that had survived the hand-rearing process to become robust juveniles. We read with amusement reports from our colleagues in Cape Town telling us how the young birds reacted with some confusion upon being liberated from their transport boxes onto the beach. Because they had been raised by humans, instead of racing directly toward the ocean as the adult penguins had done, they gathered in front of the rescue workers and gazed up at them, apparently confused as to what they were supposed to do next. Several even wheeled around and waddled right back into the boxes that had been used to carry them to the coast and to freedom. At least the boxes and the people were something familiar. But with some gentle prodding from their surrogate parents, they finally turned toward the sea and, upon spying their adult counterparts, scurried over and followed them into the cold ocean waters for the very first time.

This had been the largest rescue of oiled penguins–or of any group of animals–in history, and due to the tremendous international response, most of the penguins affected by the *Treasure* oil spill survived their traumatic ordeal. Of the 19,000 oiled penguins that were rescued, 90 percent survived and were released back into the wild; and of the 38,500 that were handled (including both the 19,000 penguins that were oiled and the 19,500 that were relocated), 95 percent survived! Of all the adult penguins directly affected by the *Treasure* oil spill, just 1,868 lost their lives. Of those, 241 were lost in the move to Cape Recife and another 965 had to be euthanized at the rescue centers because they were too sick or too emaciated to save. It's remarkable that, with so many thousands of penguins rescued, such a small percentage died. Miraculous, really. I don't think that any of us, upon first walking into Salt River or SANCCOB and seeing the masses of oiled penguins packed into the holding pens, could have ever imagined such a positive result.

Sadly, the toll on the chicks was far greater. Of the estimated 15,000 penguin chicks born that season, about 4,000 died after their parents were covered with oil or relocated to Cape Recife. Of the estimated 7,000 older chicks that were in the last stages of fledging and left on the islands, most are assumed to have survived. Of the 3,350 chicks that were taken in for hand-rearing, 2,287 survived to fledging and were released. Given the known survival rates of young

African penguins, between 900 and 1,400 of all of these fledglings (both hand-reared and parent-reared) are assumed to have survived to breeding age. While the *Treasure* oil spill tragically cut short the best breeding season ever recorded for this vulnerable species, at least this group of chicks likely lived long enough to reproduce and help propagate the species.

Not only was this the largest penguin rescue ever undertaken, it was also the most successful. The survival rate was even greater than that normally achieved with oiled African penguins. (On average, SANCCOB's success rate is about 75 percent.) But none of it could have taken place without the experienced leadership and unfaltering commitment of rescue organizations such as SANCCOB, IFAW, and IBRRC, or without the dedicated help from researchers, conservation scientists, and field staff from a host of local organizations. More important, this record-breaking rescue would never have succeeded without the tireless efforts of more than 12,500 steadfast and determined volunteers. The number of penguins they saved during the *Treasure* rescue is equal to half of the entire population of the species today. Without these selfless volunteers, the species would be one step closer to extinction.

20

The Lucky Survivors—How Did They Fare?

The fact that we are likely to come through this disaster losing less than one percent of the world African penguin population and not forty percent [as projected] is due to the amazing and committed response of a vast number of people and organizations.

—LES UNDERHILL, AVIAN DEMOGRAPHY UNIT, UNIVERSITY OF CAPE TOWN

Scientists, as well as laypeople, naturally have many questions about the long-term repercussions of oiling on penguins, and some individuals question whether rehabilitating oiled wildlife is a worthwhile endeavor at all. They argue that most oiled birds don't survive the rehabilitation process, and even if they do, they often die shortly after being released. They also contend that there's no value in saving them, because many birds fail to reproduce successfully after being oiled. And, if the animals that are injured are not an endangered species, they don't believe there's any reason to try to help them. Some opponents of wildlife rescue efforts argue that rehabilitators are just bleeding-hearts and misguided idealists who can't stand the sight of a suffering animal, and that their efforts are completely wasted. Personally, I don't understand how anyone could walk by an injured or oiled animal and not feel their heart tearing out of their chest. It is our very humanity that compels us to help any creature in distress. Even more important, whether that animal is an endangered species or not, alleviating their suffering is simply the right thing to do.

While it's true that some bird species don't fare very well after being oiled, the gloomy statistics these skeptics and opponents generally refer to are outdated–oftentimes more than thirty years old–and don't reflect the success rates achieved since more recent developments in rehabilitation techniques and strategies. With a timely and proper response, these rates (including post-release survival rates) average 50 percent to 80 percent for most bird species. This certainly makes rehabilitation efforts worthwhile. Moreover, penguins are unique in their ability to cope with handling and the presence of humans in general, and it's been proven that these resilient seabirds do remarkably well, both during rehabilitation and afterward.

In the years since the *Apollo Sea* and *Treasure* oil spills, researchers in South Africa have closely examined whether oil ingestion, the stress of being handled daily, the disruption of being separated from their mates, and the effects of being housed indoors for long periods of time in overcrowded conditions had any negative impact on the reproductive success and overall health of the penguins. Field data gathered on the penguins oiled in both of these spills have shown that the impact is surprisingly low. While the breeding success of oiled penguins is slightly lower (only 11 percent less) than that of never-oiled penguins, their survival rates are the same, which means they will continue to reproduce for as many years as their unoiled counterparts, making a vital contribution to the overall population. Already classified as a vulnerable species at the time of the *Treasure* oil spill, the African penguin population has experienced a steep and alarming decline in the last ten years due to several environmental factors. Still, their reproductive success rate after being oiled is higher than for any other bird species in the world, validating the worthiness of rehabilitation efforts for this species in particular.

African penguins have paid a terrible price by living in the midst of a major shipping lane. In the last fifteen years, shocking numbers of these penguins have been oiled as a result of the increasing proliferation of cargo ships and tankers in their home waters. In the forty-two-year period between August 1952 and May 1994, approximately 15,000 African penguins were caught in oil spills, an average of 357 penguins per year. In just one six-year period–between the *Apollo Sea* disaster in June 1994 and the *Treasure* oil spill in June 2000–more than 60,000 penguins were directly affected. This averages out to a staggering 10,000 penguins per year. Fortunately, SANCCOB has the experience and the international partnerships in place to deal with major disasters; funding, however, is harder to come by. Before SANCCOB was established in 1968, virtually every oiled penguin died. Due to the organization's dedicated efforts over the last forty years, more than 85,000 penguins and other seabirds have been spared horrible deaths by oiling.

And, luckily, penguins' hardy nature gives them the ability to endure the stress of rehabilitation, making them excellent candidates for future rescue efforts. Their high survival rate following rehabilitation also bodes well for them, as these seabirds will most certainly continue to be subjected to oiling, due to the steady stream of aging vessels passing through their nesting and foraging grounds. Whether it's another large spill or just the constant small purges of oil from passing ships, African penguins–and their rescuers–will have to contend with this deadly problem for a long time to come.

Several penguins in the Cape Town region have been through the rescue and rehabilitation process three or four times now, and–fortunately or unfortunately–some seem to have adjusted to it. We witnessed this phenomenon with the penguins that had survived the *Apollo Sea* spill only to be oiled again in the *Treasure* spill. Although six years had passed between the two events, the previously oiled birds learned how to take food from us much faster than their never-oiled counterparts, presumably because they remembered doing it before. They also seemed less nervous in general, and more resigned to the process. (Almost as if they were thinking, "Oh ya–*this* again.") This remarkable memory and adaptability certainly helped one African penguin in particular when it was mysteriously oiled again two years after the *Treasure* oil spill. A research team, led by Bruce Dyer from MCM, was on Robben Island gathering data on new polymer wing bands that were being field-tested, when they came across an emaciated penguin that was heavily coated with oil. They approached it cautiously, not wanting to scare it off. The penguin, however, just stood there watching the humans nonchalantly as they moved closer. It never tried to run away and it didn't even struggle as Bruce picked it up and put it into a transport box. It wasn't that the bird was too weak to fight; it had enough strength, but it just didn't seem that concerned. They deduced that this penguin had been through at least one rescue before and knew what to expect, as it didn't show the usual signs of alarm or distress at being approached or handled. It was almost as if the bird had been standing there, waiting patiently for someone to come along and rescue it.

Later that evening, after tube-feeding the penguin some fluids to rehydrate it, the bird surprised everyone by doing something else quite out of character for a wild penguin. As Bruce was pointing at it, the penguin reached over and began nibbling gently on his outstretched finger. This wasn't the usual painful chomp inflicted by a frightened penguin, but rather a tender investigation of his hand. Undoubtedly, the bird was extremely hungry, and must have remembered that human hands were sometimes the source of a much needed meal. Upon inspecting the ID number on its wing band, they discovered that this penguin had indeed been rescued and rehabilitated during the *Treasure* oil spill.

In a paper written by Jenny Griffin (Bruce's field assistant at the time of this unusual encounter), she and the team expressed surprise at the bird's apparent memory of being rescued and fed by humans. Having worked daily with penguins for close to nine years, I'm not at all surprised that this bird remembered its earlier experience, and that it associated humans with food, and possibly with help. The penguins that I've worked with at the New England Aquarium still recognize me when I stop by for a visit, even though it's been nearly six years since I worked there. When they hear my voice they immediately start calling and displaying, and many come swimming across the exhibit or racing across their island to greet me. Why shouldn't this penguin have remembered its human caretakers as well?

Although more than 12,500 people devoted weeks or months of their lives to saving the penguins oiled during the *Treasure* disaster, only a handful of them have had the opportunity to see the long-term result of their efforts firsthand. Our teammate Lauren DuBois was one of those lucky few. Two years after the rescue effort, she returned to South Africa to participate in a research project monitoring the African penguin population on Robben Island. She worked alongside an Earthwatch team led by Dr. Peter Barham from the University of Bristol. Dr. Rob Crawford, Professor Les Underhill, Dr. Phil Whittington, and Mario Leshoro were also on the island with them–all had played vitally important roles in the *Treasure* rescue effort and all were investigators on this Earthwatch project.

As they drove across the small island to the study site, Lauren thought back to the last time she had been on Robben Island. It was during a release of the orphaned chicks she and Steve had raised at Salt River, and at that time, the beaches had been completely devoid of penguins. Of course, all of the penguins that should have been there were either huddled in holding pens at one of the rescue centers, waiting to be washed, or partway through their long swim home from Cape Recife. Now, as they approached the same rock-strewn beach where she had last seen the chicks that she had fed and fretted over for four long weeks, Lauren was startled to see that the shore was teeming with penguins. "*This* is what it's supposed to look like," she thought to herself. She was thrilled to see so many penguins there, and it suddenly struck her that some of these were probably "*her*" penguins–the ones she had taken care of two years earlier during the oil spill.

One of her tasks on the current project was to record the ID numbers on the metal wing bands the birds were wearing. Penguins that had been banded during the *Treasure* oil spill had a distinguishing prefix number on their ID bracelets, and researchers were particularly interested in observing these individuals so they could learn more about their survival rates and reproductive success after being oiled and rehabilitated. It was early evening, which was a good time to watch for penguins, as they would be returning from their daily foraging trips at sea. Sitting alone on the rocky beach with her spotting scope, the reflection from the setting sun dancing on the surface of the water in Table Bay, Lauren began to see some penguins come ashore with the designated *Treasure* prefix number on their bands. Her excitement mounted as more and more of the *Treasure* birds appeared. When Lauren described this stirring moment to me many years later, the emotions she experienced that evening long ago came bubbling to the surface once again.

"Fortunately, there were numerous times that I was in an area doing spotting by myself," she told me, "because I would just start tearing up, going, 'Oh my *God*–these birds are *here*. These are the twenty thousand penguins that we took care of! They're coming back; they've just been out to sea, they're feeding, they're heading back to their nest, or they're hauling out for the night.' And I remember sitting there thinking, 'My God. How often in your life do you get the chance to say you did something right–that you did something that made a real difference? And this made a difference, because these penguins were *here*. It was two years later and they were still doing well; they still looked healthy. It was incredible." She went on to say that this experience helped her to resolve some of the complex emotions she had been burdened with since leaving South Africa two years earlier. She also was able to reconcile the sense of unfinished business she'd had following the rescue effort. "When we left South Africa after the oil spill, we didn't really know how the birds were going to do. We didn't have that closure. When we all left we thought, 'I *think* they're going to be okay.' But we really didn't know. You need a little more of an ending than that, and this trip was a bit of closure for me. It was really nice to have that–to know I had actually done some good, and I could now move on to the next chapter." By the time she finished recounting her story, we both were weeping.

Obviously, the breeding season in the year 2000 was severely impacted by the oil spill and the ensuing rescue effort. Although the season had already peaked by the time the penguins were rescued, some birds returned to their islands immediately upon their release from the rescue centers or from Cape Recife, and resumed breeding again, presumably to replace their lost eggs or chicks. But there had been a tremendous amount of disruption–pairs that were separated during the rescue had to try to find each other, reestablish their pair bonds, and reclaim their nesting sites. Because it was the tail end of the season, most birds did not attempt to breed again that year, and researchers speculated that some separated pairs never located each other again. This meant they would have to start off the next breeding season by finding new partners. In the year following the *Treasure* oil spill, however, the reproductive success of the previously oiled penguins rivaled that of their unoiled counterparts. In fact, 2001 was a particularly successful breeding season for both oiled and unoiled penguins, due to the influx of huge schools of pilchards and anchovies in the region that year. While fewer penguins bred overall, ostensibly because of pair disruption and disturbances to their annual molts (possibly from being housed indoors for so long), those that did breed were very successful at raising their chicks to fledging.

Research following the earlier *Apollo Sea* spill had shown that previously oiled penguins were not quite as successful as their unoiled counterparts at rearing chicks in years when food availability was poor, at least in the first two years following rehabilitation. Unfortunately, the steep reduction in the population of anchovies and pilchards in the western Cape region has since become a severe problem for the penguins, dramatically reducing the breeding success and overall survival of both groups of birds. Although formerly oiled penguins have similar incubation success (raising eggs to hatching) as unoiled penguins, their chicks grow at slightly slower rates, and they seem to lose more chicks as they near fledging. While the exact reason why they tend to lose older chicks is not yet understood, it seems that some previously oiled adults are simply unable to meet the needs of large, hungry offspring as they demand more and more food.

And, though the reproductive success of females oiled in the *Treasure* spill is slightly lower than that of unaffected females, it is still very good. Researchers have estimated that 74 percent of the female penguins oiled in this spill continued to breed successfully, whereas 85 percent of never-oiled females experience breeding success. It is also important to note that penguins oiled in other disasters have had even better breeding success than the ones that were oiled during the *Treasure* oil spill. The birds that were rehabilitated during the *Treasure* rescue effort were held longer than usual, due to the immense number rescued at once. Researchers believe that both the length of time the penguins sat covered with oil before being washed, and the overall length of their stay in the rescue centers, may have had an impact on the reproductive success of these individuals. These particular findings underscore the need to move birds through the system as quickly as possible during future rescue efforts.

Following the successful relocation of the 19,500 penguins to Cape Recife during the *Treasure* rescue, researchers proposed utilizing the same innovative strategy in future catastrophic spills, the idea being to focus initial efforts on removing clean penguins from the area as quickly as possible before they become oiled, then relocate them and let them swim back to their colonies once the oil has been cleaned up from the area. While it doesn't prevent all the birds in a given area from getting oiled, this approach has many advantages: it requires far less manpower, the logistics are less complicated, and it's much less expensive than rehabilitating oil-covered animals. Additionally, because the reproductive success of never-oiled penguins is higher than that of oiled penguins that have been rehabilitated, this approach would yield the best results in terms of population growth.

And there is some encouraging news. An unexpected finding from the follow-up studies of the *Treasure* birds was the significant survival rate for the chicks that had been hand-reared by humans. In the wild, only 15 percent of juvenile African penguins survive to breeding age; however, the abandoned chicks that were rescued during the *Treasure* oil spill appear to have had a higher survival rate following their release. The success of the penguins raised by humans might have been due to having a steady food supply during the hand-rearing process. Although, there was actually a problem with chicks being overfed at one of the offsite rearing facilities; some chicks were possibly lost because of this, and others were so portly when they were released that researchers worried they might not be able to maneuver quickly enough to catch fish and evade predators.

Penguin parents aren't always able to provide enough food for their growing chicks, but this was not an issue for the chicks raised at the rescue centers. These chicks also had about an extra month of nurturing before having to hunt for their own food, because they were released at an older age than the age at which parent-reared chicks are generally kicked out of the nest. In addition to being held back until they were nice and robust, these chicks had the added benefit of multiple swimming sessions prior to being released. Although young penguins instinctively know how to swim when entering the ocean for the first time, they are initially quite awkward, and it takes some time before they become skilled swimmers. It has been speculated that the enforced swimming practice at the rehabilitation centers gave the rescued chicks an advantage over their parent-reared counterparts, as they would have gained both strength and experience before being released, making capturing prey and escaping from predators more likely during their early days at sea.

Once these hand-raised chicks reached breeding age (generally at about four years old, although at least one of the *Treasure* chicks bred successfully when it was not quite three years old), they appeared to be just as successful–if not more so–at raising offspring than their naturally raised counterparts. It is estimated that out of 1,000 hand-reared female penguin chicks, about 135 survive to breeding age, and those 135 birds generate about 1,220 chicks over the course of their lifetimes. Given this evidence that the hand-reared chicks are successfully raising their own young and contributing to the overall population of the species, it is vital that efforts are made to raise abandoned chicks whenever possible.

The positive outcome from hand-rearing the abandoned chicks during both the *Treasure* and the *Apollo Sea* oil spills has changed the thinking and strategy of conservation scientists in boosting declining penguin populations. Given the survival rate and breeding success of the human-raised chicks, hand-rearing could prove to be part of the solution to the current population crisis experienced by several penguin species around the globe. In South Africa, hundreds of chicks are abandoned near the end of each breeding season, when fish move away from the penguins' feeding grounds or when their parents begin to molt earlier than usual. Once the adults start dropping their feathers, they can no longer go to sea to capture prey for themselves and their chicks. And, in recent years, African penguins have had to swim much farther than usual from their rookeries to find food, often keeping them away too long to sustain their growing chicks. Where their foraging trips used to last a day or less, they now take up to three days. Undoubtedly, the current shortage of pilchards and anchovies in the Western Cape region has added undue pressure to the breeding birds, and with each passing year, the number of abandoned and starving chicks rises.

But there is hope: a concerted effort is being made by SANCCOB to rescue and raise these youngsters. Shortly after the *Treasure* oil spill, SANCCOB (with support from IFAW, Bristol Zoo Gardens, the ADU, CNC, MCM, Dyer Island Conservation Trust, Robben Island Museum, and the New England Aquarium) established the Chick Bolstering Project. Since 2001, they have taken in and raised close to two thousand orphaned chicks. Between 2006 and 2008 alone, they admitted 1,424 chicks, 1,194 of which they successfully reared to fledging. In a recent interview, Venessa Strauss, current CEO of SANCCOB, stressed the soundness and value of this innovative conservation project. She stated: "Our research shows that hand-rearing African penguins has a significant effect on conserving the wild population, with hand-reared [. . .] chicks showing higher survivorship to breeding age and higher productivity than birds that fledge naturally in the wild. This makes each individual chick very precious to our efforts to conserve this already vulnerable species."

More than 8 percent of the chicks they have rescued over the last nine years have been released back into the wild as healthy juveniles. Without SANCCOB's intervention, all of these chicks would have perished out on their islands. South Africa and Namibia's penguins are already teetering on the edge of extinction. The Chick Bolstering Project is an incredible opportunity—perhaps one of our last—to save the species.*

* See Appendix II for information on how to donate to this project.

21

Life After the Rescue—
The Mission Continues

You gain strength, courage and confidence by every experience in which you really stop and look fear in the face. You are able to say to yourself, "I have lived through this horror. I can take the next thing that comes along." You must do the thing you think you cannot do.

—ELEANOR ROOSEVELT

With the exception of losing my mother due to a careless medical mistake in June 2004, my participation in the *Treasure* oil spill rescue was both the most difficult and the most significant experience of my life. And I imagine it will always remain so. It was one of the most grueling things I've endured physically, mentally, emotionally, and spiritually. Yet, at the same time, it was a tremendously rewarding and profound experience, not only because I had the awesome privilege of helping to save a species threatened with extinction–as I had always dreamed of doing–but because I had the rare opportunity to be part of something so much bigger than myself. An astonishing number of people–all strangers to each other–had come together and given of themselves unreservedly in an effort to accomplish something vitally important and seemingly impossible. And the most incredible thing was that we were so successful–nearly 40,000 penguins were saved from this environmental disaster.

It was awe-inspiring and extremely moving to see how quickly thousands of people from around the world mobilized to help the traumatized penguins. There was a tangible sense that something extraordinary was happening at the two rescue centers–something verging on the miraculous. Never before had so many people joined forces to save a species in distress. The rescue centers were imbued with an atmosphere of hope and purpose, love and community. Both volunteers and rescue staff were often heard saying that the overwhelming volunteer response had restored their faith in mankind.

One of the more remarkable things about the volunteers was their infinite willingness to do the most strenuous, most menial, and most exhausting of tasks–all without hesitation or complaint. Most of the people taking care of the penguins at Salt River and SANCCOB–99 percent of them–did not have to be there. They were not wildlife rescuers or penguin experts or veterinarians. And while a few had also helped during the *Apollo Sea* rescue, the majority had no previous experience handling animals, let alone wrestling wild penguins. Most of them had no idea what they were signing up for when they showed up at the crowded, chaotic, stinking rescue centers for the first time. And surely, they all had other things they could have been doing with their time and their energy. But they came because they could not turn away from the penguins in their most desperate hour of need. Their love of animals and nature drew them to the rescue centers to help–and it kept them there until the crisis was over and the last penguin had been released back into the wild.

The work we did during those weeks–and, for some rescuers, months–was relentless, backbreaking, and emotionally punishing. The sight of thousands of weak, starving, shell-shocked penguins covered with oil was one of the most heartbreaking things I've ever witnessed; every minute of the experience was brutal. My teammate, Martin Vince, recently said of his time in Cape Town, "Leaving South Africa was like finishing a grueling college exam: you felt proud of the achievement, but you would be perfectly happy not to repeat the experience." And another teammate, Steve Sarro, echoed his sentiments. "This experience was one of the highlights of my life," he said, "both the most incredible adventure and yet the most heinous. I'm glad to have been a part of it, yet I *never* want a repeat performance." Still, hard as it was, I too am profoundly grateful that I had the opportunity to be a part of this rescue; in the end, it was worth the exhaustion and stress, as well as the bites, scratches, and many scars that still mark our bodies. And I would do it again in a heartbeat if the need ever arose. Of course, I hope and pray that it never will. All of my teammates have voiced the same sentiment–to a person, they've said that, despite the hardships, if there was ever another oil spill they'd be on the next plane to go help. Truly, how could one not?

When I later asked Anton Wolfaardt what the experience had been like for him on a personal level, his eloquent response echoed the feelings of everyone I've spoken with who was fortunate enough to have taken part in this rescue. "I remember occasionally thinking that it was amazing how many people–from all over the world–had taken an interest in the spill and the plight of the penguins," he said. "It was really inspirational, and had a very positive influence on others. There was an amazing sense of camaraderie, passion, commitment, and common purpose that must have contributed hugely to the final success of the rescue operation. Everyone was willing to go beyond the extra mile to do what was necessary. I felt proud and privileged to have been part of the exercise, and to work with so many amazing people. I know that my colleagues at Cape Nature Conservation feel the same way, and still talk about it as one of their most memorable experiences to this day. What made the experience so rewarding for those involved is that we could see the tangible outcomes of our efforts so quickly–such as birds being cleaned and released. You really felt like you were making a difference."

This is an experience that will stay with me forever. When I'm old and frail, I imagine that as I look back on my life, the *Treasure* rescue will stand out as one of the most meaningful experiences I ever had. As a child, I believed that I was meant to do something significant to help animals. After returning from South Africa, I remember feeling as though I had just met and fulfilled that destiny. I also recall thinking with great certainty that if I died the next day it would be okay, because I had contributed to something that mattered–to something that truly made a difference. While this may sound grandiose or a bit melodramatic, it was actually a very humbling moment. I knew, having done this, I could die happy.

They say we aren't given more than we can handle. I learned during this rescue operation that I could handle so much more than I had ever thought possible. When I first arrived in South Africa and heard what my assignments would be, I wasn't sure I would be able to rise to the occasion. But, somehow, I did. During that period in Cape Town, I discovered a deep well of strength, resolve, and confidence that I had not been completely aware of before. This experience opened my eyes to what I was truly made of and what I could actually achieve. I had been tested, and was surprised to find that, when pushed beyond the edge of all I knew, there was a source from which I could still draw. This was an empowering realization, and I returned from South Africa stronger and more self-assured than I had been just a few weeks earlier. And I knew beyond the shadow of a doubt that I could now handle anything else that life might throw at me–I was sure nothing could be harder than this rescue had been. I remember expressing this to several friends at the time, not knowing how soon I would need that newfound strength.

Four years later, I found myself nursing my mother through the harrowing four-month hospitalization that marked the end of her life. The internal resilience I had discovered in South Africa helped get me through the horror and heartache of that traumatic time. I am ever grateful for my participation in the *Treasure* rescue effort for many reasons; one is that it prepared me–in a way that nothing else could–to be fully present for my mother; and it enabled me to support her in a very meaningful way at the end of her life.

The rescue effort was life-changing and extraordinary in so many ways, but it was the people who left the most indelible impression. The indefatigable volunteers were what made the whole experience not only tolerable but uplifting. And the mere fact that so many of them showed up to help the penguins still makes me shake my head in wonder. Shortly after returning from South Africa, I gave my first public presentation about the *Treasure* oil spill. My very wise, insightful friend and mentor, Susan Brown, was in attendance, and at the conclusion of the program she raised her hand to ask a question.

"Why," she asked, "do you think so many people came to volunteer?"

"To help save this vulnerable species from extinction," I answered.

"No," she said. Then, measuring her words slowly–in the way she often does when she wants me to stop and really think carefully about a question before answering it–she asked me again, "What *really* brought them there?"

After pondering for several moments I realized what it was–what she already knew.

"Love," I said. "It was love that brought them there."

The love of more than 12,000 people saved those 38,500 penguins, and my hope is that the love of even more people will help ensure a future for all penguins, and for all animals. That's why I wanted to share this story. When I decided to leave my job at the New England Aquarium, I wasn't sure what would come next professionally, but I knew that I wanted to continue teaching others about penguins and to be an advocate for them. From that vision, my company, The Penguin Lady, was born in 2006. Since that time, as an educational consultant and lecturer, I've been writing about penguins and sharing my passion for them with audiences in the United States and abroad. As "the Penguin Lady," I have a mission to raise both awareness and funding to help protect these engaging seabirds. To that end, I have donated part of the proceeds from every appearance to penguin rescue, research, and conservation groups. A portion of my proceeds from the sale of this book will be donated to these groups as well.

To learn more about the many challenges penguins are now facing, and to find out how you can help them, please see the Epilogue and Appendices (Appendix II has a list of organizations that need funding to continue their important work of protecting penguins. Penguins need our help now more than ever. While mankind has historically been their biggest threat, we are also now their only hope. These delightful birds have waddled their way into our collective hearts. We now have the opportunity and responsibility to protect them.

It would have been impossible for just one person to save the thousands of penguins affected by the *Treasure* oil spill; but in coming together, an enormous community of caring people did. The same holds true for efforts to protect and preserve all penguin species. If every person commits to doing a little bit, we just might be able to save them. As Margaret Mead once said, "Never doubt that a small group of thoughtful, committed citizens can change the world. Indeed, it is the only thing that ever has." I know this to be true, for I have lived it. The book you hold in your hands tells but one story of how a group of people did just that. Together, we *can* make a difference.

EPILOGUE:

An Uncertain Future—The Forecast for Penguins

Humankind has not woven the web of life. We are but one thread within it. Whatever we do to the web, we do to ourselves. All things are bound together. All things connect.

—TED PERRY, SCREENWRITER,
(OFTEN ATTRIBUTED TO CHIEF SEATTLE)

Birds are indicators of the environment. If they are in trouble, we know we'll soon be in trouble.

—ROGER TORY PETERSON,
NATURALIST AND ORNITHOLOGIST

On January 15, 2010, the unthinkable happened. An oil slick suddenly appeared in Table Bay, not far from where the *Treasure* had gone down almost ten years earlier. The slick, which eventually covered an area a quarter of a mile long, was believed to have come from the iron-ore carrier decaying on the ocean floor 165 feet below. Over time, powerful underwater currents and the corrosive forces of salt water had broken down and shifted the remains of the *Treasure,* and hidden pockets of oil still trapped on the vessel were released into the surrounding waters. When news of the oil leaking into the bay reached me 8,000 miles away in Boston, my immediate reaction was, "No! No! This cannot be happening!" The thought of the penguins getting oiled for a second time from the same shipwreck was devastating. And, with the future survival of the species already in question, the timing of this subsequent spill from the *Treasure* could not have been worse.

The drifting oil soon threatened to infiltrate the saltwater intake for the Koeberg nuclear power plant's cooling system, putting the station on high alert. The plant raised its underwater booms and, after a few tense days, the slick moved away from the immediate area. Media reports only mentioned the penguins incidentally, and I spent several anxious days waiting to hear if they had been affected. One week later, SANCCOB posted an update on its website stating that conservation workers on the penguins' breeding islands had not yet reported seeing any oiled penguins. After another two weeks, the reports were the same. This was puzzling, as the oil slick was hovering directly in the midst of their feeding grounds, but it was a relief, nonetheless. The penguins, it seemed, had dodged a bullet. But the incident was a sober reminder of the continued threats from these shipwrecks. It is estimated that, worldwide, there are more than 3 million vessels–from massive oil tankers to small personal yachts–littering the ocean floor. What future dangers to wildlife do all of these wrecks pose?

This recent oil slick eerily echoed a similar episode that had taken place just nine months earlier in Namibia, though in that event, the penguins were not lucky enough to escape the oil. On April 8, 2009, oiled African penguins started appearing in their breeding colonies near Lüderitz. It began with a small trickle of affected birds, but within a few days, 171 oil-covered penguins were spotted on Mercury, Ichaboe, Halifax, and Possession Islands. These four islands alone support 96 percent of the Namibian population of the species. As there had been no reports of a ship being damaged or sinking in the area, the source of the oil–which appeared to be bunker fuel–was a mystery. The oil spill, which ultimately polluted 150 miles of coastline, was the largest the country had ever seen. It was suspected that the oil had seeped from the corroding hull of the *Meob Bay*, a fishing trawler that sank near Lüderitz in June 2002. As in so many shipping disasters, no salvage operation was ever conducted to remove the oil from the vessel after it went down. And despite the fact that the Namibian population of African penguins is highly endangered, no rescue centers exist there. So the Ministry of Fisheries and Marine Resources in Lüderitz took on the task of rescuing and rehabilitating the oiled penguins. The staff managed to wash all of the penguins, but they weren't truly prepared to conduct a large-scale or long-term rehabilitation effort. Lacking the proper space, resources, and manpower to care for so many animals, they were soon overwhelmed, and called upon SANCCOB for help.

On April 20, Venessa Strauss, CEO of SANCCOB, flew to Namibia, where she obtained a flatbed truck, packed up 129 of the strongest penguins, and drove non-stop back to the rehabilitation center in Cape Town, a trip of 900 miles that took twenty solid hours. Our former teammates, Gayle Sirpenski and Steve Sarro, just happened to be in South Africa at the time for the 2nd International African Penguin Conference, so they went to SANCCOB, suited up in oilskins, and helped take care of the rescued birds. Their experience, this time, was far less stressful. After caring for the penguins for one month, SANCCOB staff and volunteers released the birds on Derdesteen Beach, a few miles north of Milnerton Beach, where more than 17,000 penguins had been liberated during the *Treasure* oil spill rescue. As with the *Treasure* releases, large crowds of concerned Cape Town residents gathered to watch and celebrate as healthy penguins streamed out of their transport boxes onto the sandy beach. Drawn to each other like magnets, the anxious birds immediately flocked together, forming one large group before rushing en masse toward the open ocean. It marked the first time that African penguins from Namibia were released in South African waters, and there was no guarantee that the birds would return to their native islands hundreds of miles to the north. But, based on Peter, Pamela, and Percy's long swim home from Cape Recife nine years earlier, rescuers and researchers felt confident that these penguins would find their way home again as well. And they did.

The first sighting of one of the penguins back on its breeding island was on the evening of June 8, 2009. This penguin, easily identifiable by the pink spot freshly spray-painted on its white breast, had completed its historic journey from Cape Town to Mercury Island–a distance of about 650 miles–in just over two weeks. Within a few days, other pink-breasted penguins began turning up on their islands as well. More than 96 percent of the penguins survived the rehabilitation process; just six birds had died. The rescue had been a success.

Late in 2009, yet another grounded iron-ore carrier threatened the penguins in Table Bay. On September 7, the Panamanian-registered ship *Seli 1* ran aground in Cape Town. The twenty-nine-year-old vessel came to rest on Sunset Beach, directly in front of Big Mike's office in Blouberg. Shortly after the ship grounded, Mike sent me photos taken through his office window of the disabled ship stuck fast in the sand just a few hundred feet away. In the last six years, at least four bulk carriers have run aground in Table Bay. Fortunately, in each case, a swift salvage response prevented thousands of tons of oil from spilling into the bay. But the threat of another disaster is ever present. Every day, approximately eight commercial ships enter Cape Town Harbour, and dozens of other cargo carriers and tankers pass by. With so many ships regularly plying the rough seas near Cape Town, the likelihood of future oiling incidents looms large. The disturbing reality is that another spill is practically inevitable; it's only a matter of time before another aging vessel meets its end in these waters and, once again, thousands of penguins are oiled.

Although there haven't been any major oil spills near Cape Town since the sinking of the *Treasure,* SANCCOB routinely rescues and rehabilitates approximately 1,000 oil-soaked penguins every year. Why, then, do these penguins continue to get oiled despite there being no recorded spills in the region? There are several reasons, including increased ship traffic around the Cape of Good Hope; corruption within the shipping industry; illegal dumping of bilge and ballast water and other unreported spills; the placement of major shipping lanes and busy commercial ports in the midst of the penguins' breeding and foraging grounds; and the continued use of an aging fleet of vessels in notoriously rough seas.

Since the *Treasure* oil spill, conservation scientists and wildlife advocacy groups have intensified their efforts to protect these vulnerable seabirds, and the South African government has been doing its part as well; some progress has been made, but efforts must continue. Thus far, changes have included establishing no-fishing zones near some of the penguins' breeding islands; designating marine protected areas along other parts of the South African coast; and, in some locations, adding artificial nest boxes to the penguins' breeding grounds. The good news is that these conservation strategies seem to be making a difference. The Dyer Island Conservation Trust (DICT) recently began installing nest boxes in several rookeries, which the penguins have been utilizing successfully. Since there isn't an appropriate substrate for the penguins to dig into to make a burrow (because all the guano was removed from their islands during decades of guano harvesting), these sturdy fiberglass shelters provide the birds with much-needed protection from predators, storms, and overheating in the sun.*

The recent establishment of marine protected areas is also showing evidence of being quite effective. In 2009, Marine and Coastal Management instituted a no-fishing zone around St. Croix Island, in hopes of protecting the world's largest breeding colony of African penguins. To determine the efficacy of banning fishing within 15 miles of the island, several months prior to establishing the ban, researchers with Cape Town University's Percy FitzPatrick Institute put satellite tracking devices on some of the St. Croix penguins to monitor their foraging trips. After the area was closed to fishing, researchers once again tracked the penguins' foraging trips using satellite tags. In all, ninety-one penguins were studied.

In February 2010, researchers reported that the fishing restrictions had been extremely beneficial for the penguins. They discovered that, after fishing was prohibited near the island, the penguins

---

* See Appendix II for information on how to support this project.

made much shorter trips in search of food. In fact, they spent 30 percent less time and 40 percent less energy hunting. Before the ban, penguins spent an average of two days swimming up to 100 miles in search of food, with 75 percent of their feeding dives occurring more than 15 miles away from their breeding island. After the ban, 70 percent of their dives were made within the 15-mile no-fishing zone around their island, a complete reversal in both ratio and behavior. Several penguins on Bird Island, which sits in the same bay as St. Croix Island, were also satellite-tagged as control subjects. During the same period, the area around their island remained open to fishermen. They found that the foraging behavior of this group of penguins did not change and, interestingly, they spent even more energy than before in search of prey.

Although this study did not occur during their breeding season, researchers anticipate that more efficient hunting trips will boost the penguins' reproductive success during future breeding seasons. With shorter foraging trips, the parents can return to their nests more frequently with bellies full of fish, and fewer chicks should starve. And because the parents will not have to expend as much energy finding food, their chances of survival should increase as well. Hopefully, the encouraging findings from this study will be enough to convince authorities to establish similar marine protected zones around every penguin breeding island. These protected areas are but one step in protecting these charismatic seabirds; however, more work still needs to be done to ensure their future survival, particularly for the rapidly shrinking Namibian population.

The International Fund for Animal Welfare and several other conservation groups are also working hard to establish, change, and enforce shipping laws and regulations to protect penguins and other animals harmed by frequent oiling. Their goals are to have fines increased for oil spills, make shipowners and insurers accountable for preventable shipping accidents, increase patrolling of local waters to prevent ships from dumping contaminated ballast and bilge water, and establish even more marine protected areas. They also hope to get the shipping lanes moved, as conservation biologist Dr.

Dee Boersma has done in Argentina. For the last twenty-five years, Boersma has been studying Magellanic penguins, a species that has long been afflicted by the same chronic oiling that is harming African penguins (to whom they are closely related). Until 1997, shipping lanes off the coast of Argentina overlapped with the migration route of Magellanic penguins, and every year, an estimated 40,000 penguins were killed from oil pollution. By satellite-tagging the penguins, Dr. Boersma proved that the ships and the penguins navigated the same waters. Working in collaboration with the Wildlife Conservation Society, she then convinced authorities to move the shipping lanes, thus saving the lives of many thousands of penguins. Moving the heavily traveled shipping lanes away from the African penguins' foraging grounds and breeding islands may prove to be more problematic, however, due to the positioning of major shipping ports near these sensitive areas. In other efforts to protect the species, scientists have discussed changing the conservation status of African penguins to endangered. But will it be too little, too late?

Frightening figures have begun to emerge from studies conducted in the ten years since the *Treasure* oil spill. Just before this ship sank, the world population of African penguins was estimated to be 50,000 breeding pairs. It is now half that. Between 2000 and 2006 alone—just a short six-year span—the total population fell by 40 percent to 31,000 pairs. And between 2006 and 2009, it fell even further. In April 2009, more than eighty penguin researchers, rehabilitation experts, conservation scientists, and penguin aquarists (including Steve and Gayle) gathered in Gansbaai, South Africa, for the 2nd International African Penguin Conference to discuss the status of the species. A few months later, based on what was disclosed at the meeting, BirdLife South Africa issued a press release. It said, in part:

> *In the 1920s, despite more than a century of sustained persecution, principally from egg collecting and guano scraping, around one million pairs of African Penguins (Spheniscus demersus) bred at Dassen Island, off the West Coast of South Africa. Now the global population is a mere 28,000 pairs. As for Dassen, last year fewer than 6,000 pairs nested. That's half*

> *a percent of the former numbers. Averaged out over 100 years,* this collapse represents a loss of 20,000 birds per year from just one colony; equivalent to 1,600 birds a week, or more than two birds per hour. *This phenomenon is not unique to Dassen Island, but is an example of the massive reduction in African penguin numbers around our coast.*

Put into these vivid terms, the decline of this species is truly alarming.

Since this conference, their numbers have decreased yet again. There are now just over 25,000 breeding pairs. Prior to the *Treasure* oil spill in June 2000, there were 17,181 breeding pairs of African penguins on Dassen Island and 5,700 on Robben Island. Today, there are 5,100 and 2,400 pairs, respectively. The exact reason for this sudden and precipitous crash is not yet fully understood, but it appears that large numbers of penguins are starving to death due to the decrease of pilchards and anchovies in the Western Cape region. Some of this regional desertion may be part of a natural fifty-year cycle in an eastward movement of these fish stocks, but much of it is likely due to human activities, as well as competition with Cape fur seals for food and territory.

It is deeply disturbing to realize that many of the penguins we worked so hard to save were undoubtedly victims of this recent population crash. To imagine thousands of those rescued penguins later starving to death is heartbreaking. These feisty birds had managed to survive one man-made disaster, only to succumb years later to another, more insidious one. The rate at which they are disappearing is not sustainable. If this negative population trend is not halted, African penguins could be extinct by 2020.

And, sadly, the African penguin is not alone in its potential death march. During the twentieth century, the populations of most penguin species plummeted by 90 percent. *Ninety percent.* And populations are still dropping rapidly. Some species have only 1,000 or 2,000 individuals left. At this rate, several penguin species may very well go extinct within the next few decades. These dramatic population declines can be directly attributed to human disturbance–principally through the invasion and destruction of their habitat. It began during the mid-nineteenth century when several penguin species were exploited for their eggs and their guano–hundreds of thousands of penguins were even boiled down for their oil–and the pressure continues today through a number of practices that are harmful to penguins and to their fragile marine ecosystem. These robust and highly adaptable creatures have survived and thrived on this planet for 60 million years, yet in the blink of a cosmic eye, we've nearly wiped them out. The moment human beings realized that penguins–and the food that they eat–could be profitable, their future survival was in jeopardy. Now their very existence hangs in the balance.

The International Union for Conservation of Nature, in collaboration with field researchers and conservation scientists, has been closely monitoring penguin populations for many years, and its assessment regarding the outlook for these popular seabirds should shake some of us out of our complacency. According to the IUCN's Red List of Threatened Species, at the present time, thirteen of the planet's eighteen penguin species are at risk of extinction; four are listed as "Endangered," nine as "Vulnerable" or "Near-Threatened," and only five as "Least Concern," indicating their populations are widespread, abundant, and currently not threatened. Alarmed by the steep reduction in their numbers during recent years, a number of research scientists and conservation organizations are petitioning to have all eighteen species declared "Endangered."*

The future survival of all penguins is now at stake. We have to cease all activities that are harmful to them, and start reversing some of the damage that has been done. Penguins are being threatened by global warming, pollution, introduced predators, and irresponsible fishing practices. We are destroying our oceans–*their* oceans–at an unprecedented rate; we are removing fish in such massive quantities through overfishing and as bycatch that, in the last fifty years, 90 percent of the large, predatory fish in the oceans have disappeared. Every year, millions of tons of unwanted fish, marine mammals, sea turtles, and seabirds are unintentionally caught in the enormous nets

* See Appendix I for more information on the decline of other penguin species.

used by commercial fishing vessels. All of these unwanted animals are dumped back overboard, dead or dying. A mere fraction of what is hauled up from the ocean's depths is kept.

Several species of fish that people routinely eat have already gone extinct, and scientists recently predicted that, by the year 2048, almost all edible seafood will have been removed from our oceans. When one form of plant or animal life is removed from an ecosystem, the intricate web that is the binding force of that unique system begins to unravel and, eventually, the whole ecosystem collapses. In her recently published book, *The World Is Blue,* renowned oceanographer Dr. Sylvia Earle explains that the ocean's fate and ours are inextricably linked, and she warns that we cannot continue to harm the oceans without ultimately harming ourselves. Earle suggests that one step each of us can take to save our oceans is to remove seafood from our diets. This may sound radical, but scientific evidence points to the validity of her argument. This would protect both the fish that are hunted for consumption and those that are caught incidentally. Not to mention that some of the marine creatures we eat (and that millions of pets and livestock eat in their food) are primary prey items of penguins–in particular, anchovies, sardines, squid, and krill. If we eliminate the penguins' food, it won't be long before penguins disappear as well.

Penguins are captivating and amusing creatures–and the fact that they are disappearing is upsetting to many animal lovers–but we should be concerned about their rapidly shrinking populations for another, even more important reason: they are an indicator species (the proverbial "canary in the coal mine"), alerting us to the deteriorating health of our oceans. When the population of an indicator species spirals downward, it's a red flag warning us that that animal's environment is in crisis and we need to pay very close attention to what is happening in that ecosystem. It is a warning we cannot afford to ignore. Alarm bells should be going off for all of us, not only because of vanishing penguin species but also because of the estimated 75–150 unique plant and animal species that go extinct every single day. Although penguins are highly visible victims of the mistakes we have made as stewards of the earth, these charismatic creatures are merely symbolic of larger issues lurking beneath the surface. All life is connected. If we want to save penguins, we have to protect the planet, which requires changing our behavior. We must make this a personal commitment, a collective mission, and a global priority. Ultimately, we will be saving ourselves.

We must become vigilant about the health of the ocean, and change the indifferent attitude that we–as individuals, as corporations, and as societies–have had for so many years. For far too long, our oceans have been regarded as free dumping grounds for our personal and industrial waste; as an endless source of fish, crustaceans, and marine mammals to be removed for our consumption; and as places that take a backseat to outer space in terms of exploration, funding, and our understanding of them. Although human beings have arbitrarily delineated five main oceans, in reality, there is only one ocean, which covers 71 percent of our planet. That ocean is dying, and we must put it first–not last. This call to action is for anyone who cares about *any* living creature, from the largest animals on earth–the majestic Blue whales–to the swarming masses of tiny pink krill these gentle giants feed on. The ocean is the source of all life as we know it, and the health of the planet is entirely dependent upon the health of the ocean. It's not too late to fix the mistakes of the past. We have the ability to turn things around, and to heal this enormous and wondrous ecosystem that covers most of our planet. As Arthur C. Clarke sagely pointed out: "How inappropriate to call this planet Earth, when clearly it is Ocean."

Acknowledgments

It is no easy task to properly thank everyone who has had a hand in bringing this book to life. Undoubtedly, I will forget some people who, in one way or another, played a part. For that, I offer my sincerest apologies ahead of time. I must start by thanking my amazing parents for their unwavering love and support, and for teaching me the importance of giving back. They instilled in me a love of reading and an appreciation of nature from a very early age. My father and I had a tradition of watching nature shows together, and we must have seen every National Geographic and Jacques Cousteau special that ever aired. And I am quite certain that my mother, who was always my biggest cheerleader, helped orchestrate the birth of this book from beyond.

There aren't enough words to express my gratitude to my rockin' agent, Julie Barer of Barer Literary. Julie's brilliant guidance and unbridled enthusiasm for this project are deeply appreciated. Every author should be so fortunate to work with someone like her. And I would be remiss if I did not thank author Bret Anthony Johnston for introducing me to Julie at Grub Street's Muse & the Marketplace writer's conference. A series of wonderful coincidences (or possibly my mother) brought the three of us together that weekend. Many thanks go to my editor at Free Press, Wylie O'Sullivan, for also being so enthusiastic about this project. When she told me she had cried openly on the subway while reading my proposal, I knew I had found the right home for this book. And many thanks to Wylie's editorial assistant, Sydney Tanigawa, for all of her help. I am profoundly grateful to the entire team at Free Press for their belief in me and in this book. In particular, I must thank Dominick Anfuso, Martha Levin, Carisa Hays, Laura Cooke, Christine Donnelly, and Suzanne Donahue.

Sincere thanks go to Lori Glazer for planting the seed that led me to share this story, and to editor and author Brando Skyhorse, who had asked me to pen a different book about penguins. When the story of the oil spill rescue kept coming out instead, he graciously encouraged me to find a home for the book that I really wanted to write. Thanks also to friends and writers Kate Victory Hannisian and Alvin Powell for their expert advice during the early stages of this project. I would like to extend my gratitude to the wonderful librarians at the Topsfield Town Library for constantly cheering me on as I wrote. Special thanks go to librarian Wendy Thatcher for occasionally grabbing me by the collar and insisting I take a walk with her on her lunch break. And I don't know where I would be without my brilliantly talented friend and faithful writing companion, Donna Childs. I am indebted to her for being there across the library table from me every Saturday for months on end. I hope to return the favor while she completes her books.

I am grateful to my big brother, Fred deNapoli, for taking on more of the care of our father while I was fully immersed in the writing of this book. And many thanks to gifted author and fellow animal lover, Sy Montgomery, for being so supportive and encouraging as I navigated the later stages of it. A true kindred spirit, she has been incredibly helpful and inspiring. Many thanks to the most loving, supportive, and faithful group of friends one could ever ask for: the amazing Chipper's Gang. They are my second family. Thanks to all of my friends and former colleagues from the New England Aquarium for always inspiring, teaching, and entertaining me. They are the most dedicated, talented, brilliant, passionate, funny, and adventurous people I have ever had the pleasure of working with. And I am very grateful to Sandy and Dick Glessner, Phi Theta Kappa honor society advisors extraordinaire, who believed in me so many years ago, and who saw potential that I was not yet fully aware of. Under their tutelage, I finally overcame my deadly fear of public speaking, and even came to enjoy it. Without this skill, I would not have been hired at the New England Aquarium, and I would have missed the extraordinary opportunity to participate in this historic rescue effort.

I am deeply indebted to everyone who shared their stories, memories, and experiences from the *Treasure* rescue. In particular, I would like to thank Estelle van der Merwe, Mariette Hopley, Mike Herbig, Anton Wolfaardt, Phil Whittington, Dianne De Villiers, Gary Egrie, Didi Ettisch, Kara Masachi, Rick Meijer, and Roland Jolink. Sadly, just as I was finishing this book, Roland succumbed to ALS. I am ever grateful that, before he passed away, we were able to reconnect via Rick to talk about our shared experiences during the rescue effort. My deepest appreciation goes to each of my teammates that I shared this unforgettable experience with: Steven Sarro, Lauren DuBois, Gayle Sirpenski, Martin Vince, Alex Waier, Jill Cox, and Heather Urquhart. Heartfelt thanks go to Jay Holcomb, Linda Elliott, and Sarah Scarth for leading with so much heart, resolve, and commitment. Karen Trendler, Sam Petersen, and everyone from IFAW, IBRRC, and SANCCOB deserve special recognition for their dedicated efforts as well. Thanks also to Les Underhill, Rob Crawford, Peter Barham, and Venessa Strauss for their input, and for filling in some of the blanks. And for so generously providing photographs, I am most grateful to Les Underhill, Tony Van Dalsen, Martin Vince, Marc Dove, Jill Meyers, the DICT, and the *Cape Times*.

For sharing their love and appreciation of animals and of the magical underwater world, I am eternally grateful to Jane Goodall and Jacques Cousteau. Watching them, I learned that it was possible to carve out a career doing what one was most passionate about. They were profoundly inspirational role models when I was young, and Ms. Goodall in particular continues to be one. At seventy-six years old, she still travels more than 320 days a year, sharing her message of hope and urging us to be better stewards of this planet.

To my partner in life, Marc Dove, for being so supportive and patient through this long journey, my deepest thanks. I now have a thorough understanding (as does he) of the reason every author thanks their long-suffering loved ones for putting up with their protracted absence while they wrote their books. Marc also supported the rescue effort while I was in South Africa by collecting donations from his co-workers; thanks again to his colleagues at netNumina for contributing so generously to help the penguins.

Many thanks to the wonderful people of Cape Town, for welcoming us so warmly and for caring so much about the oiled birds. For soothing thousands of penguins during the *Treasure* rescue and afterward, Welcome–the amazing penguin whisperer–will always have a special place in my heart. And finally, to every volunteer and every rescue professional who gave so generously of themselves and worked so hard to save South Africa's penguins, my undying gratitude for helping the birds in their greatest hour of need. You are all my heroes and each of you will be in my heart forever.

APPENDIX I

The Pressure on Penguins—Current Challenges

In the end, we will conserve only what we love, we will love only what we understand, and we will understand only what we are taught.

—BABA DIOUM, SENEGALESE CONSERVATIONIST

As this book was going to press, the African penguin was officially declared an "Endangered Species" by the IVCN and BirdLife International. This news is at once both disturbing and encouraging. As illustrated in the Epilogue, however, not only are African penguins struggling to survive, nearly all penguin species are under tremendous pressure, and each is trying to scrape out an existence in an increasingly challenging environment. Of the eighteen penguin species we currently share this planet with, just five are considered to have stable populations; these are the Emperor penguin, King penguin, Adélie penguin, Chinstrap penguin, and the Little Blue penguin. Two species are listed as "Near-Threatened," meaning they will likely be classified as "Vulnerable," "Endangered," or "Critically Endangered" in the near future; the Magellanic and the Gentoo fall into this category. Six penguin species are listed as "Vulnerable," which means they are at high risk of extinction in the wild; these are the Humboldt penguin, Fiordland penguin, Macaroni penguin, Snares penguin, Royal penguin, and Southern Rockhopper penguin. Five are now listed as "Endangered," meaning they have a very high risk of extinction in the wild. Without concerted conservation efforts to protect them, many of us may witness the extinction of these four species within our lifetimes. They are the African penguin, Yellow-eyed penguin, Galápagos penguin, Erect-crested penguin, and the Northern Rockhopper penguin. Two of these–the Yellow-eyed and Galápagos–have fewer than 2,500 individuals globally, and the Galápagos penguins are barely hanging on with just 1,000 birds left. Clearly, the forecast for penguins is troubling.

A number of issues, both historic and current, have contributed to these population declines, and climate change now tops the list as a primary factor threatening the future survival of many penguin species. Global warming has brought about changes that affect these marine birds in a variety of ways: from increased ocean temperatures and shifting cold water currents (killing or displacing the fish they normally eat); to severe rainstorms that drown chicks and eggs; to the increased frequency and intensity of El Niño events. During these events, there is a major shift in the cold water currents which carry the fish and other prey that several species of penguins eat. The currents move deeper and further away from the land, making it extremely difficult for the penguins to find enough food.

Galápagos penguins, which live right on the equator in the Galápagos Islands, are particularly vulnerable to these episodes of extreme weather. In addition to being unable to find enough food to sustain themselves during El Niño years, Galápagos penguins will cease breeding. As a result, no new penguins are born to replace those that are lost due to age, illness, or starvation, further contributing to the decline in the population. These ocean-warming events have led to the starvation of thousands of penguins. In the 1982–83 El Niño, 77 percent of the entire Galápagos penguin population starved to death, and a further 66 percent of them perished in the 1997–98 El Niño. During a trip to the Galápagos Islands in 2008, I saw just a handful of these small penguins, and I wondered if it might be the first and last time I would ever see them. Just one more severe El Niño could potentially wipe out the remaining 1,000 Galápagos penguins.

Penguins living and breeding in Antarctica are also being severely affected by climate change–but in very different ways. Temperatures on the Antarctic Peninsula have risen more than anywhere else on earth: the rate of warming there is five times faster than the general rate of global warming. On some parts of the peninsula, the average temperature has risen by 5 degrees F (2.8 degrees C) in the last fifty years. These warmer temperatures have led to unusually high levels of precipitation–in the form of both rain and snow–on the peninsula. As air and ocean temperatures rise, evaporation increases. Because the air is warmer, it holds more moisture; after rising, the moisture in the air falls again as rain or snow. Antarctica has always been the driest place on earth, and until twenty-five years ago, rain had only rarely been recorded there. Now it falls regularly on the peninsula. In fact, when I was there in January 2009, I witnessed a rookery full of Gentoo penguin chicks shivering violently in a cold, spitting rain. No longer able to fit entirely under their parents, they had become saturated and could not warm themselves. The chicks' downy feathers are not designed to protect them under these conditions, and many soaked penguin chicks have frozen to death after temperatures plummeted following rainstorms on the peninsula.

This increased precipitation impedes breeding success in other ways as well. There has been an increase in severe snowstorms on the Antarctic Peninsula, with snows so deep that penguins are literally buried as they sit on their nests. Photos taken during these storms show hundreds of lumps beneath the snow blanketing the rookery, with a few beaks poking out here and there. Like the rain now falling regularly on the peninsula, these deep snows are a newer phenomenon, and the penguins' strong parenting instincts tell them to stay put to protect their eggs and chicks, no matter what. Evolution did not prepare them to deal with this new challenge. Researchers have found colonies of mummified penguins that had clearly perished after being buried alive in a sudden blizzard. Even if the penguins survive a major storm, when the deep snows melt, their chicks and eggs are susceptible to drowning, or they can get soaked in the pooling water and then freeze as temperatures drop again.

Emperor penguins are affected by rising temperatures in yet another way. This species breeds on ice shelves that form (from seawater) during the winter breeding season; as the penguins are raising their chicks, these ice shelves slowly melt back toward their rookeries. If the chicks have not fledged into their waterproof feathers by the time the ice shelves melt as far as the breeding grounds, the young chicks are forced into the ocean before they are physically ready, and cannot survive. Their thick, downy feathers become waterlogged, and they drown or freeze to death in the icy Antarctic waters. Some of these ice shelves are not growing as large as they used to in the past, and they are now melting faster and earlier than normal as well. In fact, seven ice shelves along the peninsula have disintegrated or retreated in the last two decades.

This early melting and decrease in the formation of the ice shelves affects another creature that is at once critical to the penguins' survival and an integral part of the food chain for many animals there: Antarctic krill. These small, shrimplike animals are dependent upon the sea ice, and the decrease in this ice has greatly impacted their life cycle. Antarctic krill get their nourishment from eating the algae that grows underneath the sea ice, and these vast ice shelves serve as their nurseries. Sea ice now covers 40 percent less of the West Antarctic Peninsula than it did twenty-five years ago. This reduction in sea ice has caused a reduction in algae, which, in turn, has led to a vast reduction in krill. It has been estimated that the current krill population in Antarctica is 80 percent smaller than it was in the 1970s. Between the decrease in krill due to climate change, and the decrease due to massive harvesting of these crustaceans by commercial fisheries, this food staple for penguins is rapidly shrinking. These various changes are now forcing penguins out of their established breeding grounds (many of which they have occupied for tens of thousands of years). In search of colder weather and more suitable nesting grounds, a number of penguin colonies have started migrating farther south on the Antarctic Peninsula.

In addition to the impacts of global warming, penguins have endured tremendous stress as the result of a number of other environmental abuses. These could fill a second book, but briefly, they include overfishing of their food sources, encroachment on their nesting habitat through coastal development, introduction of non-native predators against which they have no natural defenses, fishing net entanglement in both commercial and small-scale fisheries, poisoning from chemical waste and pesticides such as PCBs, DDT, and HCBs (now being found in the tissues of penguins in the remote and once pristine Antarctic), and persistent oiling events. And, in recent years, penguins have been succumbing to a variety of bacterial and viral diseases, including avian malaria and Newcastle's disease. During every breeding season in New Zealand since 2003, entire rookeries of highly endangered Yellow-eyed penguin chicks have been wiped out from mysterious diseases that have yet to be identified. Many of these pathogens found in penguin colonies for the first time have likely been introduced inadvertently by humans.

And while oiling is a continual problem for penguins and other marine animals, major oil spills from shipping accidents account for just a fraction of the oil that ends up in our oceans. About 37 million gallons of oil are spilled each year this way–but another 137 million gallons are dumped into the ocean each year during maintenance of ships' engines, and from the illegal purges of bilge and ballast water. Surprisingly, most of the oil that pollutes our oceans originally comes from land-based sources, and much of it is due to the illegal dumping of used engine oil from cars down sewer drains. Every year, 363 million gallons of this engine oil ends up in our waterways, and most eventually makes its way to the ocean. Oily runoff from roads is also the cause of some of this land-based pollution. Each year, another 92 million gallons of oil go up as smoke into the atmosphere and fall again as tiny solid particles into the ocean. And the cataclysmic, uncontrolled oil spill in the Gulf of Mexico, which began in April of 2010, will leave its deadly footprint on that region, threatening and killing countless marine animals for many years to come.

Large industry is certainly to blame for much of this pollution, but individuals bear some of the responsibility as well. One of the ways to help combat these problems is to examine our personal lifestyles and determine how we can each become part of the solution–be it buying a hybrid vehicle, riding a bicycle, or taking public transportation instead of driving; eating less seafood and meat; buying less "stuff"; recycling, purchasing recycled products and products with less packaging; turning off the computer or putting it in sleep mode when not in use; or simply turning down the thermostat and turning off the lights when leaving a room. Every small change in behavior counts, and all of these small changes can add up to make a real difference. Just as the unified efforts of 12,500 volunteers made a difference for the penguins harmed by the *Treasure* oil spill, the combined actions of each person reading this book can help make life better for all penguins–and for all creatures–living on earth.

Appendix II contains a list of penguin rescue and conservation groups that are in need of support, as well as a list of tips and resources for living a greener lifestyle.

APPENDIX II

How You Can Help Penguins—Rescue Groups and Resources

We forget that the water cycle and the life cycle are one.
—JACQUES COUSTEAU

As you've undoubtedly gathered by now, I'm a firm believer in the concept that one person *can* make a difference. Instead of feeling helpless and hopeless in the face of these challenges to penguins and their habitats, it is important to feel empowered and educated about how to help. There is a plethora of information available on the internet, but I have distilled some of it down for you. In this appendix you will find a list of penguin rescue, research, and conservation groups, as well as some resources for learning how to reduce your personal impact on the environment. I hope you find these useful.

You can help penguin researchers and rescue workers continue their vital conservation efforts by contributing to one or more of the organizations listed below. These groups are dedicated to the care, protection, and conservation of penguins and other seabirds, and each relies upon the generous support of donors, without whom they would not be able to continue their critically important work. Some of these organizations have fan pages on Facebook, and there are direct links to each group on my websites and blog. If you would like to make a donation to any of them, you can also do so through www.thepenguinlady.com, www.thepenguinlady.wordpress.com, or www.thegreatpenguinrescue.com. Click on the button that says "help save penguins." Large or small, every donation helps. If you have been moved by the extraordinary rescue of the penguins during the *Treasure* oil spill, and by the incredible dedication of the many organizations that save imperiled penguins, I hope you will join me in supporting these groups. Without funding, they cannot continue to operate and ease the suffering of so many animals in need.

Note that to contact any of these organizations by phone, you will need to put your country code before the number listed. These codes are available online. (If you live in the United States, the country calling code is 011–so, to call SANCCOB from the United States, you would dial 011 27 21 557 6155.)

African Penguins

SANCCOB–Southern African Foundation for the Conservation of Coastal Birds

www.sanccob.co.za/
P.O. Box 11116
Bloubergrandt 7443
South Africa
Tel: 27 21 557 6155
Email: info@sanccob.co.za
URL for donations: Go to www.sanccob.co.za/?m=11&s=1 and click on the green "donate now" button. You can "adopt" a penguin from SANCCOB or make a general donation for their care.

DICT–Dyer Island Conservation Trust

www.dict.org.za/
P.O. Box 78
Gansbaai 7220
South Africa
Tel: 27 82 907 5607
Email: info@dict.org.za
URL for donations: www.dict.org.za/shop_select_item.php

You can help increase breeding success for the penguins on Dyer Island by purchasing nest boxes. Or you can make a donation for the rehabilitation of the island's seabirds.

Penguins–Eastern Cape

www.penguin-rescue.org.za
Tel: 27 42 298 0100
Email: info@penguin-rescue.org.za
URL for donations: www.penguin-rescue.org.za/index.php?page_name=donation or www.penguin-rescue.org.za/index.php?page_name=support to adopt a penguin.

You can name a penguin that you adopt, or make a donation for their general care.

Percy FitzPatrick Institute of African Ornithology

www.fitzpatrick.uct.ac.za
Dr. Rob Little, Manager
c/o Percy FitzPatrick Institute of African Ornithology
University of Cape Town
Rondebosch 7701
Cape Town
South Africa
Tel: 27 21 650 4026
Email: rob.little@uct.ac.za
URL for donations: Go to www.uct.ac.za/dad/funddev/giving/ and in the drop-down window titled "I want to donate to the following project" select "Percy FitzPatrick Institute of African Ornithology."

PFI provides critically important satellite-tracking data for the African penguins. For more information about this conservation project, go to www.fitzpatrick.uct.ac.za/pdf/Project_AfricanPenguins.pdf.

Chick Bolstering Project

(a joint project with SANCCOB and several other organizations)
www.bristolzoo.org.uk
Bristol Zoo Gardens
Clifton,
Bristol BS8 3HA
United Kingdom
Tel: 44 117 974 7358
Email: cschwitzer@bristolzoo.org.uk (Dr. Christoph Schwitzer, project coordinator)
URL for donations: Donations can be made through the Bristol Zoo Gardens or via SANCCOB. To support this important conservation project through the Bristol Zoo Gardens, visit www.thebiggive.org.uk/project.php?project_id=6134&search=4c50281f-0536-4c9d-9a7d-37f27a0bf2b5. To donate directly through SANCCOB, go to www.sanccob.co.za/?m=11&s=1. Click on the "donate now" button and scroll down to "chick bolstering project." For more information, visit www.bristolzoo.org.uk/resources/documents/Conservation%20pages/projects/Foundation_-_SOUTH_AFRICA.pdf.

ADU–Avian Demography Unit

http://adu.org.za/
ADU, Department of Zoology
University of Cape Town
Rondebosch 7701
South Africa
Tel: 27 21 650 2423
Email: adu-info@uct.ac.za
URL for donations: www.uct.ac.za/usr/dad/dev/giving/USA_pledgeform.pdf
To make a donation to the ADU's penguin research projects, write

in on the donation form a request for your donation to be directed to Les Underhill's projects. (It is critical that you "request"–not dictate–that your donation be directed to Les's penguin research.)

Little Blue Penguins

Penguin Foundation
www.penguinfoundation.org.au
Tel: 61 3 5951 2800
Email: info@penguinfoundation.org.au
URL for donations: http://penguinfoundation.org.au/index.php?option=com_content&view=article&id=2&Itemid=2.

You can adopt a penguin or make a general donation to the rehabilitation center.

Oamaru Blue Penguin Colony
www.penguins.co.nz
Waterfront Road
Oamaru, Otago
New Zealand, 09001
Tel: 64 3433 1195
Email: obpc@penguins.co.nz
URL for donations: www.penguins.co.nz/?sponsor

You can adopt a Little Blue penguin to help support this organizations important conservation work.

Yellow-eyed Penguins

The Yellow-eyed Penguin Trust
http://yellow-eyedpenguin.org.nz/
P.O. Box 5409
Dunedin 9058
New Zealand

Tel: 64 3479 0011
Email: yeptrust@gmail.com
URL for donations: http://yellow-eyedpenguin.org.nz/get-involved/how-you-can-help/#online

You can make a general donation, or help increase the breeding success of Yellow-eyed penguins by purchasing tree saplings to restore their vanishing nesting habitat.

Katiki Point Penguin Charitable Trust
www.penguins.org.nz
Moeraki Lighthouse
RD 2 Palmerston
Otago 9482
New Zealand
Tel: 64 3439 4033
Email: rosaliegoldsworthy@gmail.com
URL for donations: www.penguins.org.nz/index.phtml?page_id=833;816;816;833;867

Your donation will help support the care of injured and ailing Yellow-eyed, Little Blue, and Fiordland penguins.

Galápagos Penguins

Galápagos Conservation Trust
www.savegalapagos.org
5 Derby Street
London W1J 7AB
United Kingdom
Tel: 44 20 7629 5049
Email: gct@gct.org
URL for donations: https://secure.gct.org/sponsorpenguin.html

You can adopt an endangered Galápagos penguin through this organization.

Galápagos Conservancy
www.galapagos.org
11150 Fairfax Blvd, Suite 408
Fairfax, VA 22030
United States
Tel: 703-383-0077
Email: comments@galapagos.org
URL for donations: www.galapagos.org/membership2/member ship_standard.cfm

Your donation will help support the conservancy's important conservation work in the Galápagos Islands.

Magellanic Penguins

The Penguin Project
http://mesh.biology.washington.edu/penguinProject/home
Kincaid 24
Box 351800
University of Washington
Seattle, WA 98195
United States
Tel: 206-616-2791
Email: penguin_update@u.washington.edu
URL for donations: http://mesh.biology.washington.edu/penguin Project/donate

Help Dr. Boersma save the highly threatened population of Magellanic penguins in Argentina.

Penguins of the Falkland Islands

Falklands Conservation
www.falklandsconservation.com
1 Princes Avenue

London N3 2DA
United Kingdom
Email: info@conservation.org.fk (or ann@falklands-nature.demon.co.uk)
URL for donations: same as Web URL–click on "how you can help" button on left side

You can adopt a penguin, or visit the "penguin appeal" page to make a general donation. Or visit www.justgiving.com/falklandsconservation to donate via the "JustGiving" site.

Organization for the Conservation of Penguins
www.falklands.net/
Alvear 235
Rio Gallegos
Argentina
Email: use the email contact form on the website at www.falklands.net/Contact.php
URL for donations: www.falklands.net/YouCanHelp.shtml

You can make a general contribution toward conserving the five penguin species found in the Falkland Islands, or adopt and name a Magellanic penguin (and even visit that individual penguin on your next trip to South America).

Rescue Groups for Oiled and Injured Wildlife

IBRRC–International Bird Rescue Research Center
www.ibrrc.org
San Francisco Oiled Wildlife Care and Education Center (SFBOCEC)
4269 Cordelia Road
Fairfield, CA 94534
United States
Tel: 707-207-0380
Email: no_cal_center@ibrrc.org
URL for donations: http://ibrrc.org/donate.html

IBRRC is the world's foremost authority in rehabilitating oiled wildlife. This was one of the primary groups overseeing the care of the oiled penguins during the *Treasure* oil spill. The staff there have been saving animals from oil spills worldwide since 1971.

IFAW–International Fund for Animal Welfare
www.ifaw.org
290 Summer Street
Yarmouth Port, MA 02675
United States
Tel: 800-932-4329
Email: info@ifaw.org
URL for donations: www.ifaw.org/ifaw_united_states/donate_now/index.php#x

IFAW is a world leader in the care of injured and imperiled wildlife. This organization oversaw most of the complicated logistics during the *Treasure* rescue effort; it has been rescuing and advocating for animals worldwide since 1969.

Tri-State Bird Rescue and Research
www.tristatebird.org
110 Possum Hollow Road
Newark, DE 19711
Tel: 302-737-9543
Email: Use the contact form on the website at www.tristatebird.org/contact
URL for donations: www.tristatebird.org/support/how

Established in 1976, Tri-State specializes in the rescue and rehabilitation of oiled and injured wildlife. You can support their important work by adopting a bird or becoming a member.

Other Ways to Help Penguins

Most zoos and aquariums today are actively involved in important conservation work to protect disappearing species; many facilities provide funding for field researchers and in situ conservation projects. Many institutions also provide experienced support staff during wildlife rescue efforts such as the one chronicled in this book. Check with the Association of Zoos and Aquariums (AZA) at www.aza.org; the European Association of Zoos and Aquaria (EAZA) at www.eaza.net; the World Association of Zoos and Aquariums (WAZA) at www.waza.org; or check with your local zoo or aquarium to learn about their conservation efforts, and ask them how you can support their programs. The New England Aquarium was the first aquarium in the United States to establish a dedicated Conservation Department. To learn more about the New England Aquarium's current conservation initiatives, visit www.neaq.org/conservation_and_research/index.php.

"Adopt a Penguin"

Several zoos and aquariums have "adopt a penguin" programs. Some of these programs help fund field conservation work and some help support the colony at the institution. Check with your local zoo or aquarium to see if you can adopt a penguin through them.

New England Aquarium

www.neaq.org
Central Wharf
Boston, MA 02110-3399
United States
Tel: 617-226-2162 (to adopt a penguin); 617-973-5200 (for general information)
Email: svaz@neaq.org
URL to adopt a penguin: www.neaq.org/get_involved/animal_sponsorship/proud_parent_animal_sponsorship.php.

You can become a "penguin parent" and help support the aquarium's colony of African, Rockhopper, and Little Blue penguins. The

New England Aquarium's African penguins are part of the Species Survival Plan, an AZA conservation program designed to manage and conserve species that are currently listed as "Threatened" or "Endangered."

Conservation Organizations

BirdLife South Africa
www.birdlife.org.za
P.O. Box 515
Randburg 2125
South Africa
Tel: 27 11 789 1122
Email: info@birdlife.org.za
URL for donations: www.birdlife.org.za/page/5326/overview

This organization carries out important conservation work on behalf of penguins and other bird species. You can also contact their parent organization at www.birdlife.org.

WWF-SA–World Wildlife Fund for Nature–South Africa
www.panda.org.za/
Private Bag X2
Die Boord
Stellenbosch 7613
South Africa
Tel: 27 21 888 2800
URL for donations: www.panda.org.za/?section=Act_Donate

WWF-SA does important work on behalf of penguins and many other animals in South Africa. You can also contact their parent organization at www.wwf.org.

Volunteer Opportunities with Penguins in the Field

Earthwatch Institute
www.earthwatch.org/
3 Clock Tower Place–Suite 100
Box 75
Maynard, MA 01754
United States
Tel: 800-776-0188
Email: Expeditions@earthwatch.org
URL for African penguin project: http://www.earthwatch.org/exped/barham.html.

This expedition is truly a hands-on experience. You will be working in the field with researchers who are studying the African penguin colony on Robben Island, near Cape Town. You will help scientists collect important biological data, and assist in monitoring the health and breeding success of the penguins that survived the *Treasure* oil spill.

AVIVA
www.aviva-sa.com/
P.O. Box 60573
Table View 7439
South Africa
Tel: 27 21 557 4312
Email: info@aviva-sa.com
URL to volunteer for African penguin project at SANCCOB: http://aviva-sa.com/sanccob-penguin-conservation-project-cape-town.php

You can volunteer through AVIVA for a six-, eight-, ten-, or twelve-week stint helping to rehabilitate oiled and injured African penguins at SANCCOB. This would be an amazing experience!

Greenforce
www.greenforce.org/
530 Fulham Road

London SW6 5NR
United Kingdom
Tel: 44 20 7384 3343
Email: info@greenforce.org
URL for penguin project: http://greenforce.org/expeditions/south_africa_penguin.htm

You can volunteer for a six-week stint helping to care for oiled and injured penguins at SANCCOB through Greenforce, a non-profit based in London. Their program is almost the same as AVIVA's, but Greenforce charges more, as their expedition includes tours around the Cape Town area.

Enkosini Eco Experience
www.enkosiniecoexperience.com/
P.O. Box 15355
Seattle, WA 98115
United States
Tel: 206-604-2664
Email: info@enkosini.com
URL to volunteer for their penguin conservation project at SANCCOB: http://enkosiniecoexperience.com/PenguinConservationCentre.htm

You can volunteer for a six-week stint at SANCCOB through this U.S.-based organization.

Working or Volunteering with Penguins at Zoos and Aquariums

If you are interested in becoming a penguin caretaker or aquarist, contact your local zoo or aquarium to inquire about internships, volunteer opportunities, and jobs. For a listing of accredited facilities in the United States, go to AZA's website at www.aza.org. For accredited zoos and aquariums in Europe, as well as worldwide, visit www.eaza.net (for Europe) and www.waza.org (for facilities worldwide). To supplement their staff, many facilities rely on volunteers and interns to help with the care of their penguin colonies. Interns and volunteers at the New England Aquarium assist staff with everything from cleaning to feeding to educating the public about penguins. Most zoos and aquariums hire new staff from their volunteer and intern pools. If you are considering working with penguins as a career, this is a critical first step in getting your foot in the door, and determining if it is the right career for you. Whether in the field or in a zoo or aquarium, the work is a lot harder, a lot dirtier, and a lot less glamorous than it looks.

Some Great Penguin Websites and Educational Resources

http://web.uct.ac.za/depts/stats/adu/oilspill/ (ADU's pages about the *Treasure* oil spill)

http://adu.org.za/sp003_00.php (links to the ADU's scientific papers about penguins)

http://penguinscience.com/index.php (Penguin Science–click on the "Education" button for resource materials)

www.penguinworld.com/ (penguin researcher Dr. Lloyd Spencer Davis's site, with a very thorough database)

http://penguin.net.nz/ (New Zealand Penguins–great website by penguin researcher Dave Houston)

www.seaworld.org/animal-info/info-books/penguin/index.htm (SeaWorld's penguin education pages)

www.pinguins.info/ (this is a German website, so there are a few grammatical errors in the English version, but it has very detailed and accurate information)

www.penguins.cl/index.htm (The International Penguin Conservation Work Group)

http://penguinpages.org/ (Pete & Barbara's Penguin Pages–Dr. Peter Barham's website)

Websites About the Author and Her Work with Penguins

www.thepenguinlady.com or www.thegreatpenguinrescue.com (my company website, with information about presentations, this book, and other resources)

www.thepenguinlady.wordpress.com (blog about penguins and my work with them)

www.facebook.com/pages/The-Penguin-Lady/179484222169?ref=ts (official page for The Penguin Lady on Facebook, with penguin news and educational tidbits)

http://twitter.com/thepenguinlady (The Penguin Lady on Twitter)

Resource List for Protecting the Ocean and the Atmosphere

By now, no doubt, you have heard a great deal about steps you can take to help protect our planet. We all need to take responsibility to ensure a clean and healthy environment for all animals, including ourselves. This means conserving water and energy, recycling, reducing our reliance on fossil fuels (thus reducing our carbon output), reducing our personal consumption and waste, eating seafood that is sustainable, and restricting our intake of meat. There are many online resources to learn about simple things you can do to help heal the planet–the following are just a few websites with excellent information. These groups also accept donations in support of their important shared mission.

www.oceanconservancy.org (click on "take action," then on "tips for living responsibly")

http://marinebio.org/Oceans/Conservation/ (click on "100 Ways to Make a Difference" for practical suggestions about how you can help conserve marine environments)

www.theoceanproject.org (The Ocean Project promotes worldwide ocean conservation) www.saveourseas.com (Save our Seas Foundation provides grant money for marine conservation, research, and education projects worldwide)

www.seaweb.org (SeaWeb raises awareness about the health of our marine ecosystems)

www.montereybayaquarium.org/oa/ (Monterey Bay Aquarium's website has suggestions for ways to help protect the ocean. Scroll down and click on "make smart seafood choices" for a list of sustainable seafood items)

To help reduce global warming (which will benefit penguins), use one of the carbon footprint calculators found on these websites and follow their tips for greener living:

www.epa.gov/climatechange/ (EPA's website–click on "what you can do")

www.edf.org/page.cfm?tagID=573 (Environmental Defense Action Fund's site)

www.nature.org/initiatives/climatechange/ (The Nature Conservancy's website)

Recommended Penguin Books for Adults

There are many penguin books out there, but these are a few that I think are best.

The Penguins (Bird Families of the World), by Tony D. Williams, 1995. Written and edited by penguin researchers, this is *the* definitive book about penguin biology and behavior, but it may be a bit technical for the casual reader. Sadly, it is out of print and very hard to find. Try checking your local library for a copy.

Smithsonian Q & A: Penguins: The Ultimate Question & Answer Book, by Lloyd Spencer Davis, 2007. This very engaging and informative book, written by a well-known penguin researcher, answers questions about penguins you would have never even thought to ask! A good read for adults as well as young adults.

Penguins, by Lloyd Spencer Davis and Martin Renner, 2004. Another very thorough overview of penguin behavior, ecology, and evolution by penguin researchers.

Penguins: A Worldwide Guide, by Remy Marion and Sylviane Maigret-Mondry, 1999. A nice summary of each penguin species.

The Natural History of the Antarctic Peninsula, by Sanford Moss, illust. Lucia deLeiris, 1988. The definitive guide for anyone planning a trip to view penguins and other wildlife in Antarctica.

The Adélie Penguin: Bellwether of Climate Change, by David G. Ainley, illust. Lucia deLeiris, 2002. Written by a renowned penguin researcher, this book chronicles the Adélie penguin's life cycle and its dependence upon disappearing sea ice in Antarctica. Like Tony D. Williams's book, this one is more technical in nature.

The Ferocious Summer: Adélie Penguins and the Warming of Antarctica, by Meredith Hooper, 2008. Hooper describes what she learned about the impact of global warming on this species during a season spent with penguin researchers.

Waiting to Fly: My Escapades with the Penguins of Antarctica, by Ron Naveen, 1999. A charming record of this researcher's fascination with these seabirds and his many years of studying them in the wild.

APPENDIX III

Just the Facts, Ma'am—
The Final Statistics

This is unquestionably the world's worst coastal bird disaster.

—SARAH SCARTH, IFAW EMERGENCY RELIEF DIRECTOR

The Ship–MV *Treasure*
Size: 885 ft. long x 143 ft. wide x 78 ft. deep
Owners: Universal Pearls, Piraeus, Greece
Agents: Good Faith Shipping, Piraeus, Greece
Insurers: Bureau Veritas, Paris, France
Country of registry (or flag): Panama
Age of ship: 17 years old (built in Japan in 1983)
Amount of cargo on board: 143,731 tons of iron ore
Route: The *Treasure* was traveling from Brazil to China

The Oil Spill
Date the *Treasure* sank: June 23, 2000, at 3:30 a.m.
Amount of oil spilled into Table Bay: 1,300 tons of bunker oil, 64 tons of lube oil, 56 tons of marine diesel
Amount of oil removed from the ship after it sank: 200 tons

Date the last of the known oil was removed from the sunken wreck: July 18, 2000

Date of probable second oil leak from the decaying shipwreck: January 15, 2010

Cost of cleanup and animal rescue effort (in U.S. dollars): approximately $6.1 million

Number of Oiled Adults (and Older Juveniles) Rescued

Number of oiled penguins rescued from Robben Island: 14,825

Number of oiled penguins rescued from Dassen Island: 3,516

Number of oiled penguins rescued from other locations: 413

Total number of oiled penguins rescued: 18,754

Number of oiled penguins that died in the wild before being saved: approximately 150

Number of oiled penguins that were euthanized: 965

Total number of oiled penguins that were rescued but did not survive: 1,868

Total number of oiled penguins rehabilitated and released: 16,886 (90 percent of those rescued)

Number of Clean Adults (and Older Juveniles) Rescued

Number of clean penguins rescued from Robben Island: 7,161

Number of clean penguins rescued from Dassen Island: 12,345

Total number of clean penguins transported to Cape Recife: 19,506

Number of clean penguins that did not survive relocation to Cape Recife: 241 (just over 1 percent)

Number of Adults and Juveniles Rescued (Both Oiled and Clean)

Total number of adult and juvenile penguins rescued from Robben Island: 21,986

Total number of adult and juvenile penguins rescued from Dassen Island: 15,861

Total number of oiled adult and juvenile penguins rescued from other locations: 413

Total number of adult and juvenile penguins rescued: 38,260

Total number of rescued adult and juveniles that were later released: 36,392 (95 percent)

The Penguin Chicks

Number of chicks rescued from Robben Island: 2,643

Number of chicks rescued from Dassen Island: 707

Number of chicks that were euthanized: 319

Number of hand-reared chicks that fledged and were released: 2,287 (68 percent survival rate)

Number of Chicks Born and Number Lost After Oil Spill

Number of chicks on Robben Island: about 6,000 / Number lost: about 3,000 (50 percent)

Number of chicks on Dassen Island: about 9,000 / Number lost: about 1,000 (11 percent)

Total number of chicks lost in the 2000 season due to the oil spill: about 4,000 (27 percent)

Salt River and SANCCOB

Number of penguins at Salt River: about 16,000

Number of penguins at SANCCOB: about 5,000 initially, then about 3,000 after Salt River opened on June 27 and some of the penguins were moved there

Date Salt River closed its doors (remaining birds sent to SANCCOB): August 24

Date first penguin was washed: at Salt River, July 1 / at SANCCOB, July 3

Date last penguin was washed: at Salt River, July 29 / at SANCCOB, August 10

Dates of first and last penguin releases: July 18 and October 10

Length of entire rescue effort from first capture to last release: 109 days (15½ weeks)

(To view GNLD's video *GNLD Operation Penguin* about washing the penguins, go to www.youtube.com/watch?v=IpwNbEaamRA)

The Volunteers

Number of volunteers between June 23 and August 16, 2000: about 12,500

Number of volunteers per day: up to 1,000

Number of hours donated by volunteers through August 16: more than 556,000

Number of 5-hour shifts worked through August 16: 111,200 (each volunteer worked an average of 9 shifts, for a total of 44.5 hours, although many worked much longer)

How long it would have taken one person working 8 hours per day, 7 days per week, to rehabilitate all of the oiled penguins: 69,500 days (190 years)

Value of time donated by volunteers in U.S. dollars (if paid at $20/hour): $11,120,000

The Oiled Wildlife Rescue Team

Total number of professional caretakers on Oiled Wildlife Rescue Team: about 150

Size of Oiled Wildlife Rescue Team at any given time during the rescue: about 40

Number of penguin and bird specialists brought in from zoos and aquariums: 110

Number of facilities worldwide that sent specialists: 59 facilities in 14 countries

Number of hours worked by the Oiled Wildlife Rescue Team: approximately 50,000

Peter, Pamela, and Percy's Big Swim from Cape Recife to Cape Town

Peter was released on June 30, 2000, and returned to Robben Island on July 18, 2000. (He made the 560-mile swim in 18 days.)

Pamela/Pamelito was released on July 3, and arrived on Dassen Island on July 25. (The journey took 22 days.)

Percy was released on July 5, and returned to Dassen Island on July 20. (Percy broke the record, making the trip in just 15 days.)

Distance from Cape Recife to Robben Island: 545 miles / to Dassen Island: 577 miles

Number of hits on ADU's satellite-tracking page during their swim home: more than 100,000

View this animated map at: http://web.uct.ac.za/depts/stats/adu/oil spill/sapmap3.htm

To watch Jon Stewart's report about Peter the Hero Penguin's swim from Cape Recife go to: www.thedailyshow.com/watch/wed-july-19-2000/other-news---hero-penguin

Miscellaneous Statistics

Number of liters of LDC detergent used to wash the penguins: 7,500

Amount of pilchards the penguins ate between June 2 and August 16: 400 tons (4 million individual fish)

Cost of the donated pilchards in U.S. dollars (the fishermen were later reimbursed): $105,900

Food donated to feed the volunteers: 100,000 light meals, worth more than $42,000

Number of African penguins oiled in mystery spills every year: about 1,000

Number of penguins saved by SANCCOB since 1968: approximately 70,000 (plus another 15,000 seabirds)

Number of *Treasure* penguins that have been spotted since 2000: more than 11,000

Population Statistics for the African Penguin Over the Years

(Number of colonies: 25 island and 4 mainland colonies in South Africa and Namibia)

World population of African penguins in the year 1900: about 3 million

World population of African penguins at time of the oil spill: about 170,000

World population of African penguins in late 2009: about 80,000 (including 25,000 breeding pairs)–a decline of 53 percent since 2000 and 97.5 percent since 1900

Number of African penguins on Robben Island at time of the oil spill: about 18,000 (including 5,705 breeding pairs) / Number in 2009: about 7,700 (including 2,400 breeding pairs)–a decline of 58 percent in nine years

Number of African penguins on Dassen Island at time of the oil spill: about 55,000 (including 17,181 breeding pairs) / Number in 2009: about 16,500 (including 5,100 breeding pairs)–a decline of 70 percent in nine years

Online Resources for More Information

For more information and articles about the *Treasure* oil spill, visit the ADU's website at http://web.uct.ac.za/depts/stats/adu/oilspill/index.htm

There is an excellent documentary about the *Treasure* oil spill by Journeyman Pictures on YouTube that features many of the people in this book. Look up *Endangered Species: South Africa,* or go to www.youtube.com/watch?v=YleaF0N6TJw. A DVD of this documentary can be purchased directly from Journeyman Pictures at: www.journeyman.tv/?lid=9310

And Now, the Facts About Penguins

Penguins fall under the following taxonomic classification: Kingdom: Animalia; Phylum: Chordata; Class: Aves; Order: Sphenisciformes; Family: Spheniscidae; Genus: There are six genera, including *Aptenodytes* (the great penguins), *Pygoscelis* (the brush-tailed penguins), *Eudyptes* (the crested penguins), *Spheniscus* (the banded or jackass penguins), *Megadyptes* (the Yellow-eyed penguin is the only one in this genus), and *Eudyptula* (the Little Blue penguin is the only one in this genus).

Most scientists have long recognized seventeen penguin species, but while this book was being written, this officially changed to eighteen distinct species. It was long thought there were three subspecies of Rockhopper penguins. However, through DNA testing, Northern and Southern Rockhoppers have now been genetically identified as two separate species.

The Genus and Species Names of All Living Penguins

These are grouped by genus, and then listed by size (largest to smallest) within each genus.

Emperor penguin–*Aptenodytes forsteri*

King penguin–*Aptenodytes patagonicus* (There are two subspecies of King penguin; *Aptenodytes patagonicus patagonicus* and *Aptenodytes patagonicus halli.*)

Yellow-eyed penguin–*Megadyptes antipodes*

Gentoo penguin–*Pygoscelis papua* (There are two subspecies of Gentoo penguin; *Pygoscelis papua papua* and the smaller *Pygoscelis papua ellsworthii.*)

Chinstrap penguin–*Pygoscelis antarcticus*

Adélie penguin–*Pygoscelis adeliae*

Royal penguin–*Eudyptes schlegeli* (Some scientists believe the Royal penguin is a subspecies of the Macaroni penguin.)

Erect-crested penguin–*Eudyptes sclateri*

Snares penguin–*Eudyptes robustus*

Fiordland penguin–*Eudyptes pachyrhynchus*

Macaroni penguin–*Eudyptes chrysolophus*

Northern Rockhopper penguin–*Eudyptes moseleyi*

Southern Rockhopper penguin–*Eudyptes chrysocome* (Some scientists recognize *Eudyptes filholi* as a subspecies of *Eudyptes chrysocome.*)

Magellanic penguin–*Spheniscus magellanicus*

African penguin–*Spheniscus demersus* (formerly called Black-footed penguin)

Humboldt penguin–*Spheniscus humboldti* (sometimes referred to as Peruvian penguin)

Galápagos penguin–*Spheniscus mendiculus*

Little Blue (or Fairy) penguin–*Eudyptula minor* (There are six subspecies of Little Blue penguins, including the White-flippered penguin or *Eudyptula minor albosignata.* The other subspecies are *E. m. novaehollandiae, E. m. variabilis, E. m. iredalei, E. m. chathamensis,* and *E. m. minor.*)

Selected Bibliography

Books

Aupiais, Les and Ian Glenn, eds. *Spill: The Story of the World's Worst Coastal Bird Disaster,* Kenilworth, England: Inyati Publishing, 2000.

Earle, Sylvia. *The World Is Blue: How Our Fate and the Ocean's Are One.* Washington, D.C., National Geographic, 2009.

Hockey, Phil. *The African Penguin: A Natural History.* Cape Town: Struik Publishers, 2001.

Whittington, Phil. *The Adventures of Peter the Penguin.* University of Cape Town: Avian Demography Unit, 2001.

Chapter in Booklet

Ryan, Peter. "Estimating the Demographic Benefits of Rehabilitating Oiled African Penguins." Chapter 4 in *Rehabilitation of Oiled African Penguins: A Conservation Success Story,* edited by D. C. Nel and

P. A. Whittington. Cape Town: BirdLife South Africa and the Avian Demography Unit, University of Cape Town, 2003.

Papers Presented at Conferences

Kemper, J., et al. "Penguins Crossing Borders: TransBorder Rehabilitation of Oiled Penguins from Namibia."

Paper presented at the Effects of Oil on Wildlife Tenth International Conference, Tallinn, Estonia, October 5–9, 2009. www.eowconference09.org/wp-content/uploads/02–2-strauss-penguin.pdf.

Kuyper, S., and A. J. Williams, eds. *Proceedings of the Penguin Workshop Following the Sinking of the* Treasure *in June 2000.* University of Cape Town: Avian Demography Unit, 2004.

Moldan, Anton. "Response to the Apollo Sea Oil Spill, South Africa." Paper presented at the 1997 International Oil Spill Conference, Fort Lauderdale, Florida, April 7–10, 1997. www.iosc.org/papers/01877.pdf.

Industry Newsletters

BirdLife International. "South Africa: African Penguins Get a Champion." In *BirdLife International Africa Partnership e-bulletin,* July–Sept. 2009. www.birdlife.org/regional/africa/pdfs/africa_ebulletin_sept_09.pdf.

Wolfaardt, Anton. "Does Cleaning Oiled Seabirds Have Conservation Value? Insights from the South African Experience with African Penguins." *Ocean Orbit,* September 2008. www.itopf.com/information-services/publications/documents/Ocean08.pdf.

Scientific Journal Articles

Barham, Peter, et al. "The Efficacy of Hand-rearing Penguin Chicks: Evidence from African Penguins (*Spheniscus demersus*) orphaned in the *Treasure* oil spill in 2000." *Bird Conservation International* 18 (2008): 144–162. http://web.uct.ac.za/depts/stats/adu/pdf/bclme33_penguin%20orphans%20treasure.pdf.

Crawford, Robert J. M., et al. "The Influence of Food Availability on Breeding Success of African Penguins *Spheniscus demersus* at Robben Island, South Africa." *Biological Conservation* 132 (2006): 119–125. www.environment.gov.za/HotIssues/2008/acap/BiolgcalCnsrvtionAfrcnPnguinBrdngSccss.pdf.

Griffin, Jennifer. "Learned Survival Behavior of a Rehabilitated African Penguin." *Bird Numbers* 11, no. 2 (December 2002). http://web.uct.ac.za/depts/stats/adu/pdf/bn11_2p12.pdf.

Monfils, Rean. "The Global Risk of Marine Pollution from WWII Shipwrecks: Examples from the Seven Seas." *Ocean & Coastal Management* 49, no. 9–10 (2006): 779–788. www.seaaustralia.com/documents/The%20Global%20Risk%20of%20Marine%20Pollution%20from%20WWII%20Shipwrecks-final.pdf.

Parsons, N. J., and L. G. Underhill. "Oiled and Injured African Penguins *Spheniscus demersus* and Other Seabirds Admitted for Rehabilitation in the Western Cape, South Africa, 2001 and 2002." *African Journal of Marine Science* 27, no. 1 (2005). www.docstoc.com/docs/19870361/Oiled-and-injured-African-penguins-Spheniscus-demersus-and-other.

Pearce, David A. "Climate Change and the Microbiology of the Antarctic Peninsula Region." *Science Progress* 91, Pt. 2 (summer 2008): 203–17. http://findarticles.com/p/articles/mi_go2834/is_2_91/ai_n31591088/.

Pichegru, L., et al. "Marine No-Take Zone Rapidly Benefits Endangered Penguin." *Biology Letters,* February 10, 2010. www.cefe.cnrs.fr/esp/publis/DG/Pichegru%20et%20al%20Biol%20Let%202010.pdf.

Roulx, J. P., et al. "African Penguins *Spheniscus Demersus* Recolonise a Formerly Abandoned Nesting Locality in Namibia." *Marine Ornithology* 31 (2003): 203–205 www.marineornithology.org/PDF/31_2/31_2_203–205.pdf.

Vargas, Hernan F., et al. "Biological Effects of El Niño on the Galápagos Penguin." *Biological Conservation* 127, no. 1 (January 2006): 107–114. http://cat.inist.fr/?aModele=afficheN&cpsidt=17269854.

Whittington, Phil. "The Contribution Made by Cleaning Oiled African Penguins *Speniscus demersus* to Population Dynamics and Conservation of the Species." *Marine Ornithology* 27 (1999): 177–180. www.marineornithology.org/PDF/27/27_22.pdf.

Wolfaardt, Anton, et al. "Impact of the *Treasure* Oil Spill on African Penguins *Spheniscus demersus* at Dassen Island: Case Study of a Rescue Operation." *African Journal of Marine Science*, 30, no. 2 (September 2008): 405–419. www.informaworld.com/smpp/content~content=a918425549&db=all.

Newspaper Articles Online

Cohen, Mike. "South Africa Oil Spill: Rescue Mission Saves Penguins." *The Ledger,* July 8, 2000. http://news.google.com/newspapers?nid=1346&dat=20000708&id=BJ8sAAAAIBAJ&sjid=aP0DAAAAIBAJ&pg=5575,4619179.

du Plessis, Henri. "Diving Sunken Treasure Is Fraught with Danger." *Cape Argus,* July 10, 2000. www.iol.co.za/index.php?sf=1&set_id=1&click_id=31&art_id=ct20000710094019857T626987.

du Plessis, Henri and Jeremy Lawrence. "Divers to Plunder Sunken Treasure's Oil." *Cape Argus,* June 26, 2000. www.iol.co.za/index.php?sf=117&set_id=1&click_id=13&art_id=ct20000626102213802O400964.

Goslin, Melanie. "Plugged Oil Tanker a Ticking Time Bomb." *Cape Times,* June 25, 2000. www.iol.co.za/index.php?sf=13&set_id=1&click_id=13&art_id=ct20000625222220605O420704&singlepage=1.

Hayward, Brian. "African Penguin May Be Extinct by 2020, Warn Experts." *Weekend Post,* October 13, 2007. www.weekendpost.co.za/main/2007/10/13/news/nl07_13102007.htm.

"Jeremy Mansfield Gets a Grip on Greasy Birds." *Weekend Argus,* August 4, 2000. www.capeargus.co.za/index.php?fSectionId=3571&fArticleId=qw965373781562M521.

Johnstone, Vanessa. "Penguins Make It onto Pet Food Cans." *Cape Argus,* July 28, 2000. www.iol.co.za/index.php?sf=122&set_id=1&click_id=13&art_id=ct20000728094015 61P520237.

Joseph, Norman. "Cape Town Braces for Oil Slick." *Cape Argus,*

June 29, 2000. www.iol.co.za/index.php?sf=3&set_id=1&click_id=13&art_id=ct20000629092243829P520512.

Jowit, Juliette. "Krill Fishing Threatens the Antarctic." *The Observer,* March 23, 2008. www.guardian.co.uk/environment/2008/mar/23/fishing.food.

"Koeberg on High Alert." *The Voice of the Cape,* January 16, 2010. www.vocfm.co.za/index.php?section=news&category=sanews&article=50869.

Neff, Robert. "Flags That Hide the Dirty Truth." *Asia Times Online,* April 20, 2007. www.atimes.com/atimes/Korea/ID20Dg03.html.

"Puzzled Penguin Pete Returns Home." *The Star,* July 18, 2000. www.thestar.co.za/index.php?fSectionId=132&fArticleId=qw963916500407B262.

"Penguin Volunteers Will Stay Despite Robbery." *Saturday Star,* July 10, 2000. www.thestar.co.za/index.php?fSectionId=&fArticleId=ct20000710232503296P525789.

"Student's Invention Helps Degrease Birds." *Cape Argus,* July 14, 1999. www.iol.co.za/index.php?set_id=1&click_id=116&art_id=ct19990714102543569G530258.

Williams, Murray. "How the Treasure Met Her Fate." *Cape Argus,* June 23, 2000. www.iol.co.za/index.php?set_id=1&click_id=13&art_id=ct20000623220708722S423895.

Yeld, John. "Bubbly, Tears Flow as Penguin Clean-up Ends." *Cape Argus,* July 31, 2000. www.iol.co.za/index.php?set_id=1&click_id=13&art_id=ct20000731094020808P520941.

Websites

Autoship Systems Corporation. "South Africa Spill Worse Than at First Thought." Marine Log, June 28, 2000. www.marinelog.com/DOCS/NEWS/MMJun27.html.

Barham, Peter. "Not an April Fool's Joke." Bradclin House Portfolio. www.bradclin.com/blouberg/blouberg_activities.htm.

BirdLife International. "Save the Albatross." www.birdlife.org/seabirds/save-the-albatross.html.

BirdLife International. "Mystery Illness Threatens World's Rarest Penguin." December 22, 2004. www.birdlife.org/news/news/2004/12/yellow-eyed_penguin.html.

Callahan, Barbara. "Treasure Oil Spill Administration." IBRRC. www.ibrrc.org/admin_treasure.html.

CBS News. "Salt-Water Fish Extinction Seen By 2048." November 3, 2006. www.cbsnews.com/stories/2006/11/02/health/webmd/main2147223.shtml.

Coultas, Bruce and Elizabeth Cridland. "How SANCCOB Began: The Story of Our Founder." SANCCOB. www.sanccob.co.za/?m=2&s=2.

Crawford, Rob. "Technical Information about the Satellite Transmitters." Avian Demography Unit, University of Cape Town. http://web.uct.ac.za/depts/stats/adu/oilspill/sapmap2.htm.

Crawford, R. J. M., et al. "Initial Effects of the Treasure Oil Spill on Seabirds off Western South Africa." Avian Demography Unit, University of Cape Town. http://web.uct.ac.za/depts/stats/adu/oilspill/oilspill.htm.

CTX Center for Tankship Excellence. www.c4tx.org/ctx/job/cdb/dev/search.html.

Dyer Island Conservation Trust. "2nd International African Penguin Conference." www.dict.org.za/news.php?section=view&id=32.

Hunt, Stephen. "Mopping Up with Nature's Help." Discovery Channel, March 29, 2001. www.biomatrixgold.com/case_no1.htm.

IFAW. "Rescue the Remedy for Orphaned African Penguin Chicks." October 30, 2008. www.ifaw.org/ifaw_united_states/media_center/press_releases/10_30_2008_49814.php.

IMO Library Services. "International Shipping and World Trade: Facts and Figures." www.pdfqueen.com/html/aHR0cDovL3d3dy5pbW8ub3JnL2luY2x1ZGVzL2JsYXN0RGF0YU9ubHkuYXNwL2RhdGFfaWQ9MTM4NjUvSW50ZXJuYXRpb25hbFNoaXBwaW5nYW5kV29ybGRUcmFkZS1mYWN0c2FuZGZpZ3VyZXMucGRm.

IUCN. "The IUCN Red List of Threatened Species." www.iucnredlist.org/apps/redlist/details/144810/0.

IUCN Red List of Threatened Species. www.iucnredlist.org/apps/redlist/search.

Ocean Planet. "Oil Pollution." http://seawifs.gsfc.nasa.gov/OCEAN_PLANET/HTML/peril_oil_pollution.html.

Ultimate Africa Safaris. "Penguin Clean-up Completed." August 6, 2000. www.ultimateafrica.com/august00.htm.

Underhill, Les. "A Brief History of Penguin Oiling in South African Waters." Avian Demography Unit, University of Cape Town. http://web.uct.ac.za/depts/stats/adu/oilspill/oilhist.htm. (Source for timeline of major oil spills in chapter 4.)

Underhill, L. G., et al. "Five Years of Monitoring African Penguins *Spheniscus demersus* After the *Apollo Sea* Oil Spill: A Success Story Made Possible by Ringing." Avian Demography Unit, University of Cape Town. http://web.uct.ac.za/depts/stats/adu/species/sp003_01.htm.

Vitamin Natural. "GNLD's LDC™ Stars in Penguin Cleanup Operation!" www.vitaminatural.com/gnld/gnldcleanupldc.htm.

Walton, Marsha. "Study: Only 10 Percent of Big Ocean Fish Remain." CNN, May 14, 2003. www.cnn.com/2003/TECH/science/05/14/coolsc.disappearingfish/.

Web4water. "IFAW Appeals to Save African Penguins." July 7, 2000. http://web4water.com/news/news_story.asp?id=2917&channel=4.

Video Recording

Endangered Species. Londolozi Productions. DVD. Journeyman Pictures, 2000.

Telephone Interviews

Cox, Jill. Recorded telephone interview. August 27, 2009.

De Villiers, Dianne. Recorded telephone interview. August 25, 2009.

DuBois, Lauren. Recorded telephone interview. September 2, 2009.

Egrie, Gary. Recorded telephone interview. August 25, 2009.

Ettisch, Didi (Danielle). Recorded telephone interview. December 12, 2009.

Herbig, Mike. Recorded telephone interview. August 6, 2009, and October 6, 2009.

Hopley, Mariette. Recorded telephone interview. September 2, 2009.

Masachi, Kara. Recorded telephone interview. August 4, 2009.

Meijer, Rick. Recorded telephone interview. October 6, 2009.

van der Merwe, Estelle. Recorded telephone interview. August 27, 2009.

Whittington, Phil. Recorded telephone interview. October 6, 2009.

Email Interviews and Communiqués

Barham, Peter. Email to author. October 19, 2009.

Crawford, Robert. Email to author. March 5, 2010.

DuBois, Lauren. Email interview. August 18, 2009.

Sarro, Steven. Email interview. July 15, 2009.

Sirpenski, Gayle. Email interview. August 12, 2009.

Strauss, Venessa. Email to author. May 10, 2010.

Vince, Martin. Email interview. December 21, 2009.

Waier, Alex. Email interview. August 10, 2009.

Wolfaardt, Anton. Email interview. August 26, 2009.

How We Made This Book

In an effort to be environmentally responsible, the manuscript for this book was designed and edited in large part electronically. The text paper was manufactured using fiber sourced from responsibly managed forests, and the printing plates were recycled after use. The ink used in this book contains more than 20 percent renewable resources, including soy and other vegetable-based oils. The adhesives for the case glues are solvent free, and the case boards were created with 100 percent recycled fiber. All books that are returned to the Simon & Schuster warehouse that are not resold will be recycled.

Index

About the Author

Dyan deNapoli grew up in the small, oceanside town of Marblehead, Massachusetts. Twenty-six years after her first penguin encounter at Boston's New England Aquarium, she returned as an intern and eventually become a senior penguin aquarist. After caring for the penguins for nine exhilarating years, she founded her educational company, The Penguin Lady. She now teaches audiences worldwide about penguin biology, behavior, and conservation and donates a portion of the proceeds from every appearance to penguin rescue groups. Since 1995, she has shared her passion for penguins with approximately 250,000 people. Dyan has been the guest lecturer on nature cruises to Antarctica and the Galápagos Islands, and has also traveled to Chile, Argentina, Australia, New Zealand, and South Africa to work with, teach about, and observe penguins in the wild. She lives next to a pond on Boston's North Shore and can be found at www.thepenguinlady.com.